Segmentation and Recovery of Superquadrics

Computational Imaging and Vision

Volume 20

Segmentation and Recovery of Superquadrics

by

Aleš Jaklič

Aleš Leonardis

and

Franc Solina

Faculty of Computer and Information Science,
University of Ljubljana, Slovenia

KLUWER ACADEMIC PUBLISHERS

DORDRECHT / BOSTON / LONDON

A C.I.P. Catalogue record for this book is available from the Library of Congress.

ISBN 978-90-481-5574-3

Published by Kluwer Academic Publishers,
P.O. Box 17, 3300 AA Dordrecht, The Netherlands.

Sold and distributed in North, Central and South America
by Kluwer Academic Publishers,
101 Philip Drive, Norwell, MA 02061, U.S.A.

In all other countries, sold and distributed
by Kluwer Academic Publishers,
P.O. Box 322, 3300 AH Dordrecht, The Netherlands.

Printed on acid-free paper

To our families

Contents

List of Figures

List of Tables

Preface

In recent years, superquadrics established themselves as a popular model for representation of objects and 3D scenes in computer vision, computer graphics, and robotics. Superquadrics are a family of parametric models that cover a wide variety of smoothly changing 3D shapes which are controlled with a small number of parameters. For greater versatility the superquadric shapes can be augmented with global and local deformations. Superellipsoids, a subset of superquadrics, are volumetric models particularly suited for part-level modeling of 3D scenes which directly supports reasoning and manipulation. Especially, interpretation of range images in computer vision has been influenced by various methods for recovery of superquadric models from image data.

Readers of this book can expect to find a thorough evolution and definition of superquadric models, as well as derivations of their various geometric properties. Advantages and disadvantages of superquadrics in comparison to other volumetric models used in computer vision are addressed. Applications of superquadrics in computer vision and robotics are thoroughly discussed and in particular, the use of superquadrics for range image registration is demonstrated.

The central theme of the whole book is our method of recovery and segmentation of superquadrics from range images. The method is in essence the result of doctoral dissertations of all three authors which span a nearly ten year period. It is described in detail and compared with other methods of superquadric recovery and segmentation. Numerous examples of recovery and segmentation from range images are given.

The intended audience for this book are researchers, developers, and students of computer vision and robotics. It is assumed that the readers have a general engineering/computer science background with some familiarity of computer vision issues.

Organization of the Book

The book consists of seven chapters. This first chapter introduces the concept of part-level models for description of three-dimensional shapes and superquadrics as a special case of such models. The second chapter covers geometric properties of superquadrics. Besides the definition of superquadrics, several other useful geometric properties of superquadrics are derived. The third chapter introduces different ways of extending the expressive power of superquadrics by addition of global and/or local deformations, as well in the form of a richer parameterization (hyperquadrics). Global deformations of tapering and bending are given in detail. The fourth chapter covers the recovery of individual superquadrics from pre-segmented images. We give a survey of different recovery methods while our method of superquadric recovery from pre-segmented range images based on least-squares minimization is covered in depth. The fifth chapter is on segmentation. We define our *recover-and-select* segmentation paradigm and show how the recovery of superquadrics can be tightly integrated with segmentation to achieve their recovery without any prior segmentation of range images. The sixth chapter gives extensive experimental results of the combined superquadric model recovery and segmentation on range images. The seventh chapter is a survey of various superquadric applications in computer vision, computer graphics and robotics. In particular, we show experimental results on how superquadrics can be used for range image registration. Appendices at the end of the book contain useful tools, such as the Mathematica Code for display of superquadrics, principles of structured light range image acquisition, and pointers to the software used for the experiments shown in this book.

FRANC SOLINA

Acknowledgments

During the several years long research on superquadrics and development of the method for recovery and segmentation of superquadrics from range data the authors were supported by numerous research grants. All three of the authors spent a few years in the GRASP Laboratory at the University of Pennsylvania where they were supported by various NSF, DARPA, and ARO grants. Franc Solina thanks also the Fulbright program for his scholarship.

Their research at the University of Ljubljana was supported through several grants by the Ministry of Science and Technology of the Republic of Slovenia and by the RECCAD project under the Copernicus program of the European Union.

Aleš Jaklič would like to thank Professor Monica Partridge and the Master and Fellows of Fitzwilliam College for their support during his postdoctoral visit to the University of Cambridge, England, in 1997, where some of this book was written.

Foreword

The representation of objects by their parts has a long tradition in computer-aided design, simulation, and in cognitive psychology. Indeed, in these areas it is the dominant strategy for representing complex 3-D objects. It is absolutely clear, therefore, that part representations are excellent for many computational and cognitive tasks. What has not been so clear is how they might be useful in computer vision and robotics.

The first parts representation was suggested by Binford (Binford, 1971); this is the idea of generalized cylinders. Unfortunately, the recovery of this type of representation seems to require elaborate line grouping and reasoning, which is a difficult and largely unsolved problem. Moreover, because such descriptions are often not unique, it is unclear how they aid in object recognition.

The idea of generalized cylinders has subsequently been elaborated in two very different ways. One variation is due to Biederman (Biederman, 1985), who suggested using the Cartesian product of qualitative properties such as tapering, cross-section, etc., in order to create a qualitative taxonomy of generalized cylinders. Correct choice of these qualitative properties can make the recovery process much simpler. By using deformable superquadrics to model these qualitative properties, Dickinson, Pentland, and Rosenfeld (Dickinson et al., 1992a; Dickinson et al., 1992b) were the first to successfully use this approach in real imagery.

A second alternative to generalized cylinders was suggested by Pentland (Pentland, 1986), in which he proposed a parametric version of generalized cylinders based on deformable superquadrics. He argued that the use of a parameterized implicit function, such as the superquadric, converts the problem of recovering a description into a relatively simple numerical optimization that can be made overconstrained and therefore robust.

Although this initial formulation for superquadric recovery proved unstable, by 1987 Pentland had developed a stable method for both segmentation and fitting deformable superquadrics using the Minimum Description Length (MDL) principle in a segment-and-fit paradigm (Pentland, 1987). However, this solution was extremely slow, and it was finally Solina and Bajcsy (Bajcsy and Solina, 1987) that developed the first practical method for recovering superquadric parameters. This method immediately became (and remains to this day) the standard method for fitting superquadrics.

Since then, the idea of fitting parameterized deformable models has been extended in several directions. Perhaps the most popular extension has been to employ physics-based techniques for fitting and tracking. This approach provides a robust framework for fitting and offers the possibility for natural extension to moving, dynamic scenes.

Pentland (Pentland, 1990) was the first to take this approach, using an eigenvector analysis of the resting shape to produce a new parameterization of the superquadric deformations that is linear and orthogonal (and consequently unique). This method has been successful at solving several difficult recognition and tracking problems, such as the recognition and tracking of people (Pentland and Sclaroff, 1991).

Metaxas and Terzopoulos (Metaxas and Terzopoulos, 1991) have further extended the physics-based approach by developing a class of deformable models in which both global *and* local deformations are physics-based. The global deformations capture the salient structure of object parts, while the local deformations capture the object's details.

Biederman's qualitative representation and the parametric superquadric representation have complementary properties. The qualitative representation of part structure has proven useful for grouping regions and edges, while the parametric representations have proven useful for recovering precise descriptions of shape. It is therefore natural to try to combine the strengths of the two approaches, using one for grouping, and the other for fitting and description. This idea has been developed by Metaxas and Dickinson (Metaxas and Dickinson, 1993) and by Raja and Jain (Raja and Jain, 1994), and shows great promise.

Despite all this progress, more work remains to be done, particularly in the difficult area of segmentation. There are in principle two types of segmentation methods: "segment-then-fit," in which segmentation and fitting are only loosely connected (Gupta et al., 1989b; Pentland, 1990; Darrell et al., 1990; Ferrie et al., 1993), and "segment-*and*-fit," in which segmentation and fitting are accomplished simultaneously (Pentland, 1987; Gupta and Bajcsy, 1993; Leonardis et al., 1997; Horikoshi and Suzuki, 1993), usually using an MDL criterion. We think that this book shows that great strides towards reliable recovery of superquadric parameters for complex and articulated objects are being made.

Finally, we would like to thank Franc Solina, Aleš Leonardis, and Aleš Jaklič not only for writing this book, but also for their vital contributions to the problem of recovering parametric descriptions from complex scenes. They have been a critical force in advancing this area of vision science.

Ruzena Bajcsy
University of Pennsylvania
Philadelphia

Alex Pentland
Massachusetts Institute of Technology
Cambridge

book shows that great strides towards reliable recovery of superquadric parameters for complex and articulated objects are being made.

Finally, we would like to thank Franc Solina, Aleš Leonardis, and Aleš Jaklič not only for writing this book, but also for their vital contributions to the problem of recovering parametric descriptions from complex scenes. They have been a critical force in advancing this area of vision science.

Ruzena Bajcsy
University of Pennsylvania,
Philadelphia

Alex Pentland
Massachusetts Institute of Technology
Cambridge

Chapter 1

INTRODUCTION

The goal of computer vision in the broadest sense is to enable intelligent interaction of artificial agents with their surroundings. The means of this interaction are images of various kinds; intensity images, pairs of stereo images, range images, or even sonar data. Images, which at the sensory level consist of several hundreds or thousands of individual image elements, must in this process be encoded in a more compact fashion. The goal of this coding is not just a mere reduction of the amount of data such as is required for image compression, but a compact representation of the scene appropriate for the visual task of the agent. For any reasoning or acting on the surroundings, it is advantageous that this coding of images closely reflects the intrinsic structure of the imaged scene. Distinct objects in the scene, for example, should have distinct models of themselves in the scene representation. In this way, the labeling of individual entities, necessary for higher level reasoning, becomes possible.

It has become clear in the computer vision community that neither a complete general purpose vision system, nor a complete and general shape recovery is possible in the near future. Looking for a complete and general geometric model for all kinds and levels of shape representation is hence in vain, too. One should select a shape representation according to specific assumptions or for specific tasks.

So far, many different models have been used for modeling different aspects of objects and scenes. In general, there are models for representing 2D structures in images and models for representing 3D structures that get recovered from images. The models for representing 3D structures can be grouped into local and global models. Methods for local representation attempt to represent objects as sets of primitives such as surface patches or edges. Global methods, on the other hand, attempt to represent an object or its parts as an entity in its own coordinate system. When elements of such global models correspond to perceptual equivalents of parts, we speak of part-level models. Several part-level models are required to represent an articulated object. Part-models are sometimes referred to as models of medium granularity (Pentland, 1986). Although local models are sometimes an intermediate step towards global models, we are concerned in this book only with a particular type of global model—part-level models, and in particular, superquadrics. There are many good surveys of local models, especially surface models (Brady et al., 1985; Faugeras and Hebert, 1986; Koenderink, 1990; Besl and Jain, 1986; Bolle and Vemuri, 1991).

1.1 PART-LEVEL MODELS

A part-level shape description is important for several tasks involving spatial reasoning, object manipulation, and structural object recognition. People often resort to such part description when asked to describe natural or man-made objects (Pentland, 1986). It seems that this part-level description is used by humans to construct basic categories (Tversky and Hemenway, 1984) which are a way of organizing and interpreting the actual structure of the world. The structure of the world is not accidental, but a result of various physical processes across time. Different ways of finding the apparent structure of the world have been researched in computer vision. One way of inferring the structure of the world is to understand the processes that determined the shapes that things have. Thompson (Thompson, 1917), a pioneer of theory of shape, who undertook the analysis of biological processes in their mathematical and physical aspects, claimed that complex shapes result from deformed simple ones. People, likewise, often describe complex objects by modifying a description of a simpler, prototypical shape. For such "process description" of shape, Leyton (Leyton, 1988) proposed a process-grammar that provides a set of inference rules by which the process history of an object's shape can be determined. Another way of partitioning a shape into a small set of primitives is by applying the symmetry axis or skeleton transform (Nackman and Pizer, 1985). The symmetry axis transform, which was introduced by Blum (Blum, 1973) in the mid-1960s, decomposes a 2D shape in a unique and natural way. Yet another way of finding structure in visual data is the perceptual organization approach (Lowe, 1985). This approach, which is based on ideas of the Gestalt school in psychology (Wertheimer, 1923), attempts to discover structure through spatio-temporal regularities like least-distortion and non-accidentalness.

The usefulness of various part descriptions in computer vision depends on the specific problem at hand and on related assumptions that one can make (for example, image type, available time for processing, task). Nevertheless, one can say that part-level descriptions are generally suitable for modeling obstacles in path planning or manipulation with robotic manipulators—for object-recognition, however, they are sometimes not malleable enough to represent all necessary details and researchers are looking into extending part-level models with additional layers of details.

Several researchers have attempted to define parts as perceived by human vision in mathematical terms. Koenderink and van Doorn (Koenderink and van Doorn, 1979; Koenderink and van Doorn, 1982) defined part intersections as "parabolic lines on the surfaces of objects". Hoffman and Richards (Hoffman and Richards, 1985) refined the definition by using, instead, the "negative minima of principal curvature" which works even in case of figure/ground reversal when the perceived part structure completely changes. Such definitions of parts which are articulated in terms of analytical geometry are difficult to apply to real images because of noise and non-smooth surfaces. Computation of second order partial derivatives, which is necessary in the process of part boundary detection, requires excessive smoothing of edges and other sharp discontinuities in images.

Instead of studying part boundaries, one can approach the problem of part detection from the other end. One can try to determine parts by directly defining all possible part shapes. This is generally possible by selecting a proper parametric model. Changing the model's parameters in imposed bounds, sets the limits to the possible shapes that the model can attain. We examine next the three most influential part-level models used in computer vision; general cylinders, superquadrics, and geons.

1.1.1 GENERALIZED CYLINDERS

The first dedicated part-level models in computer vision were generalized cylinders (Binford, 1971). A generalized cylinder, sometimes called also a generalized cone, is represented by a volume obtained by sweeping a two-dimensional set or volume along an arbitrary space curve. The set may vary parametrically along the curve (Fig. 1.1).

Different parameterizations of the above definition are possible. In general, a definition of the axis and the sweeping set are required. The axis can be represented as a function of arc length s in a fixed coordinate system x, y, z

$$\mathbf{a}(s) = (x(s), y(s), z(s)) \,.$$

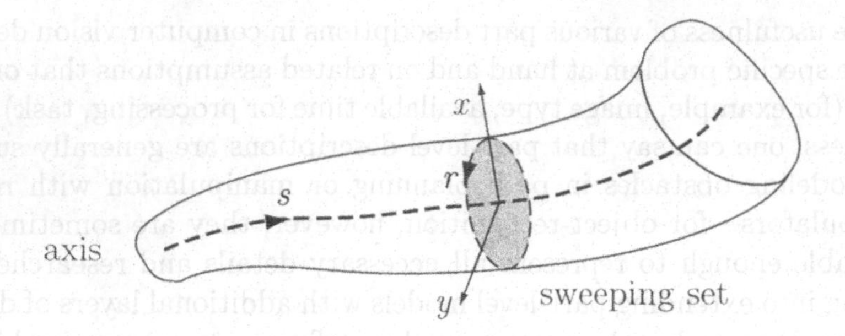

Figure 1.1. A general cylinder is defined by a cross section boundary which is translated along a spine.

The sweeping set is more conveniently defined in a local coordinate system, defined at the origin of each point of the axis $\mathbf{a}(s)$. The sweeping set can be defined by a cross section boundary, parameterized by another parameter r

$$\text{sweeping set} = (x(r,s), y(r,s)).$$

This general definition is very powerful so that a large variety of shapes can be described with it. To limit the complexity and simplify the recovery of generalized cylinder models from images, researchers often resort to restrictions. For regular objects one can use only straight axes or constant sweeping sets. Properties of straight homogeneous generalized cylinders are analyzed in detail in (Ponce et al., 1989).

Generalized cylinders influenced much of the model-based vision research in the past two decades—vision theory, as well as building of actual vision systems. Marr (Marr and Nishihara, 1978; Marr, 1982), for example, based his hierarchical part scheme for description of biological forms on cylinder-like primitives (Fig. 1.2). Generalized cylinders were also successful in vision applications. One of the often cited early vision systems, which applies general cylinders, is the ACRONYM system (Brooks, 1983). This is a rule and model based system for interpretation of airport scenes seen from the air. 2D edges that form perceptually relevant entities—defined as projections of general cylinders—are combined so that at the end, a 3D interpretation of an intensity image is obtained. Models of airplanes that the system had to recognize were assembled out of generalized cylinders.

Recovery of generalized cylinders from images was studied by many researchers. Especially notable for his research in recovery of generalized cylinders is Nevatia (Nevatia and Binford, 1977; Rao and Nevatia, 1988; Mohan and Nevatia, 1989; Ulupinar and Nevatia, 1993; Zerroug and Nevatia, 1999). In their paper (Zerroug and Nevatia, 1994) recovery and segmentation of straight homogeneous generalized cylinders from

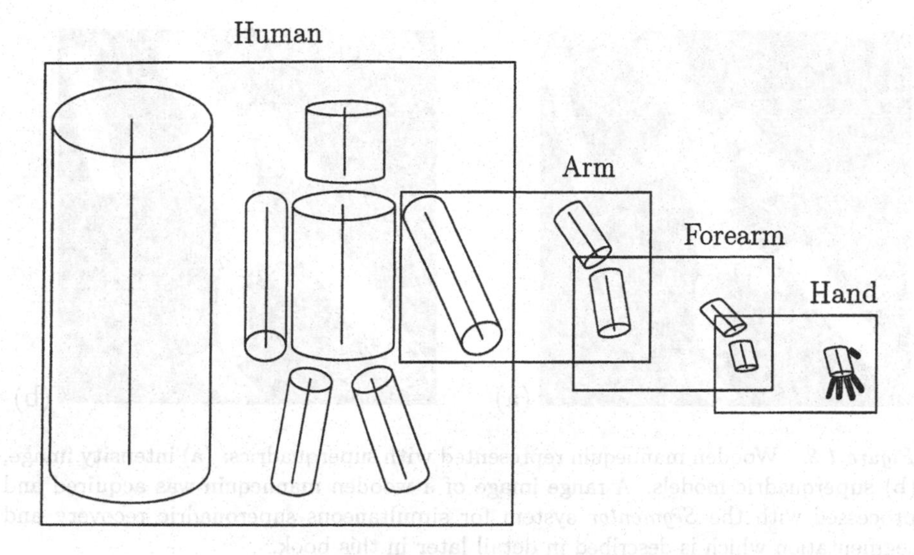

Figure 1.2. Human figure represented with generalized cylinders

real intensity images is addressed. Extruded generalized cylinders as extensions of the basic model are proposed in (O'Donnell et al., 1994). In general, the methods of recovering generalized cylinders, especially from intensity images, seem to be overly complex since they must rely on complicated rules for grouping low level image models (i.e., edges, corners, surface normals) into models of larger granularity (i.e., symmetrical contours or cross-sections) to assemble them finally into generalized cylinders. These problems are due, in part, to the complicated parameterization of generalized cylinders and to the lack of a fitting function that would enable a straightforward numerical test of how well the model fits the modeled image data.

1.1.2 SUPERQUADRICS

Superquadric models appeared in computer vision as an answer to some of the problems with generalized cylinders (Pentland, 1986). Superquadrics are solid models that can, with a fairly simple parameterization, represent a large variety of standard geometric solids, as well as smooth shapes in between. This makes them much more convenient for representing rounded, blob-like shaped parts, typical for objects formed by natural processes (Fig. 1.3).

It is interesting to note the circumstances leading to the introduction of superquadrics to computer vision. The mathematical function, at least its 2D equivalent, was first described by a French mathematician, Gabriel Lamé, in the beginning of the nineteenth century (Loria, 1910).

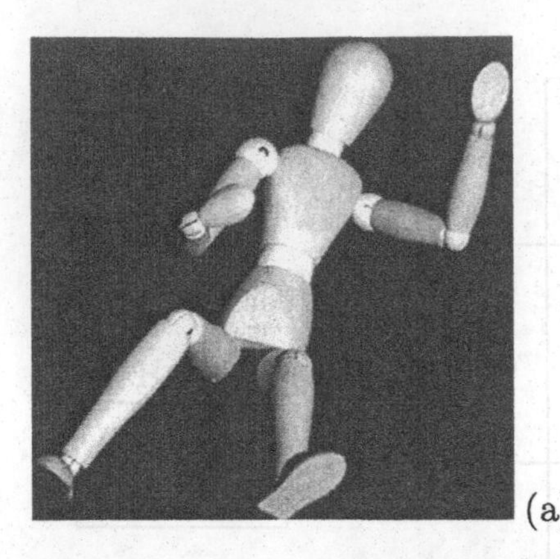
(a)

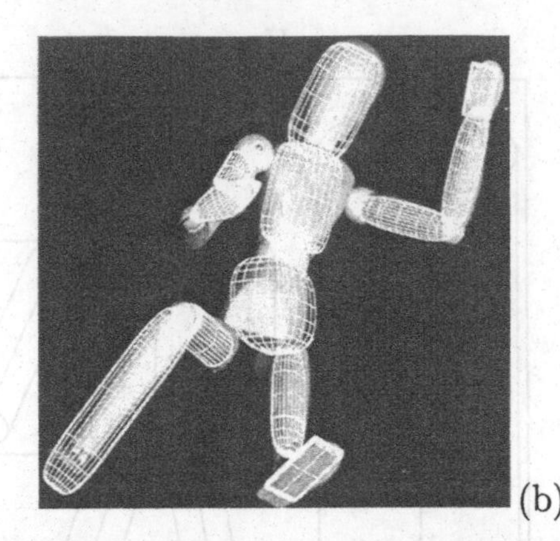
(b)

Figure 1.3. Wooden mannequin represented with superquadrics: (a) intensity image, (b) superquadric models. A range image of a wooden mannequin was acquired and processed with the *Segmentor* system for simultaneous superquadric recovery and segmentation which is described in detail later in this book.

In the 1950s, a Danish writer and inventor, Piet Hein, introduced superquadrics to architecture and design (Gardner, 1965). In 1960s, they were used for lofting in the preliminary design of aircraft fuselage (Flanagan and Hefner, 1967). In the beginning of 1980s, Alan Barr introduced them to computer graphics (Barr, 1981). Finally, a few years later, Alex Pentland brought superquadrics to the attention of the computer vision community (Pentland, 1986). A more detailed history of superquadric models is in the introduction to the next chapter.

From the very beginning, when superquadrics were introduced in computer science (Barr, 1981), the basic superquadric shape was endowed with global deformations, adding an additional layer of parameters. Even with such enriched parametric deformations, superquadrics are in fact a subset of generalized cylinders. To obtain the intuitive shapes of deformed superquadrics in the framework of generalized cylinders, a much more complicated parameterization is necessary which is, after all, not as evident as the parameterization for superquadrics.

For even more accurate modeling of shapes, different ways of introducing local deformations were proposed (Pentland and Sclaroff, 1991; Terzopoulos and Metaxas, 1991; Metaxas and Terzopoulos, 1993; Gupta and Liang, 1993; Chen et al., 1994a; Park et al., 1994; Bardinet et al., 1994; DeCarlo and Metaxas, 1998). Local deformations, which require a much larger set of parameters, can model a large variety of natural and biological shapes up to a desired accuracy.

One of the most important features of superquadric models is their interchangeable implicit and explicit defining function. The explicit form is convenient for rendering, while the implicit equation is especially suited for model recovery from images and for testing of intersection.

The suitability of a particular shape model in computer vision depends on its usefulness for recognition. It must provide a roughly unique representation which is compact, has local support, is expressive and preserves information (Brady, 1983). The final verdict, however, on actual usefulness of the chosen model in computer vision depends on how easy it is to recover the model from image data. Superquadrics are doing quite well on this test. Robust methods exist for recovery of superquadric models from pre-segmented range data (Solina and Bajcsy, 1990; Pentland and Sclaroff, 1991) as well as from non-segmented range data (Pentland, 1990; Gupta and Bajcsy, 1990; Leonardis et al., 1997).

1.1.3 GEONS

Geons are primitive building blocks for representing parts which were proposed by Biederman, in the context of human perception (Biederman, 1985). Biederman formulated a theory of recognition by components which can easily be detected and differentiated on the basis of their perceptual properties in 2D images. Geons consist of a set of solid blocks which are derived from generalized cylinders and can describe the wealth of different shapes by combining them like phonemes in a language. The set of geons consists of up to 36 primitives which were obtained by analyzing non-accidental qualitative changes on a generalized cylinder such as axis shape, cross-section shape, cross-section sweeping function and cross-section symmetry (Fig. 1.4). Biederman argued that since these differences are qualitative and reflect non-accidentalness, people should be able to classify them quickly and easily.

Geons are in essence, conceptual or meta-models of parts which are normally implemented through generalized cylinders. But geons can also be expressed in terms of superquadric models. Several methods for qualitative "Geon type" shape recovery based on superquadric models were proposed (Dickinson et al., 1992a; Raja and Jain, 1992; Wu and Levine, 1994b) which will be discussed later in the book.

1.1.4 OTHER OBJECT-LEVEL MODELS

In addition to generalized cylinders, superquadrics, and geons several other global models exist that attempt to represent an object as an entity in its own coordinate system.

Spherical harmonic surfaces are a three-dimensional analogy of Fourier components. Low order harmonic coefficients capture the gross shape characteristics (the base shape is a sphere), while the higher

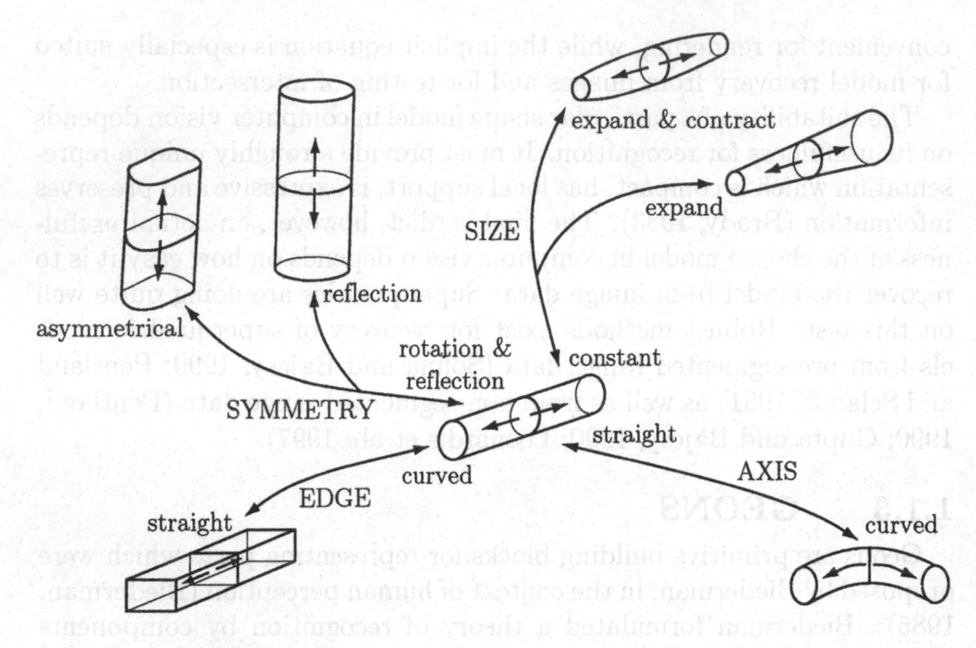

Figure 1.4. Variations over only two or three levels in the non-accidental relations of four attributes (size, symmetry, edge, axis) of generalized cylinders generate a set of 36 geon components (Biederman, 1985).

order coefficients add shape variations of higher spatial frequency (Schudy and Ballard, 1979). The principal disadvantage of spherical harmonic surfaces is that individual parameters of the model affect the entire shape in a non-intuitive way.

Gaussian images and extended Gaussian images are mappings of surface normals of a given object onto a unit sphere. In case of extended Gaussian images, a mass density, proportional to the surface area of given orientation on the modeled object is mapped onto the unit sphere. The extended Gaussian image is a unique representation for convex objects (Horn, 1986). Recently, extensions of the original concept appeared that should facilitate recovery, image registration and recognition—complex extended Gaussian images (Kang and Ikeuchi, 1993) and spherical representation (Delingette et al., 1993).

Symmetry seeking models (Terzopoulos et al., 1988b) are similar to generalized cylinders but enhanced with deformation parameters to control the elasticity of the axis and the walls of the cylinder. This physics-based modeling approach influenced also some extensions of the basic superquadric models (Metaxas and Terzopoulos,

1991; Pentland and Sclaroff, 1991), which are discussed in Chapter 3 of this book.

Blob model is an isosurface of a scalar field which is produced by a number of field generating primitives (Blinn, 1982). The basic idea is to define some points in space as sources of the potential field and then to compute the 3D isosurface of this field for a given threshold value. Blinn originally used spherical fields but superquadric fields can be used as well (Wyvill and Wyvill, 1989). Blob models of complex shapes can be recovered from range data by minimizing an energy function and recursively splitting field primitives (Muraki, 1991). The method is computationally rather expensive.

Hyperquadrics (Hanson, 1988) are a generalization of superquadrics which offer a larger shape variability through a larger number of parameters. Hyperquadrics are explained in more detail in Chapter 3.

Implicit fourth-degree polynomials (Keren et al., 1994) can represent similar shapes as superquadrics and are mathematically easier to manipulate. Fitting of implicit polynomials is discussed also in (Taubin, 1991; Taubin et al., 1994; Sullivan et al., 1994; Tasdizen et al., 1999).

The above models are better suited for representing entire complex shaped objects rather than for modeling the constituting parts of complex object in the same way as generalized cylinders or superquadrics.

1.2 SUPERQUADRICS IN COMPUTER VISION

Pentland was the first who grasped the potential of the superquadric models and parametric deformations for modeling natural shapes in the context of computer vision (Pentland, 1986). He proposed to use superquadric models, in combination with global deformations, as a set of primitives which can be molded like lumps of clay to describe the scene structure at a scale that is similar to human naive perceptual notion of *parts* and which can be recovered from images. Pentland presented several perceptual and cognitive arguments to recover the scene structure at such a part-level since people seem to make heavy use of this part structure in their perceptual interpretation of scenes. He offered superquadrics in combination with deformations as a shape vocabulary for this part-level representation. The superquadrics, which are like phonemes in this description language, can be deformed by stretching, bending, tapering or twisting, and then combined using Boolean operations to build complex objects.

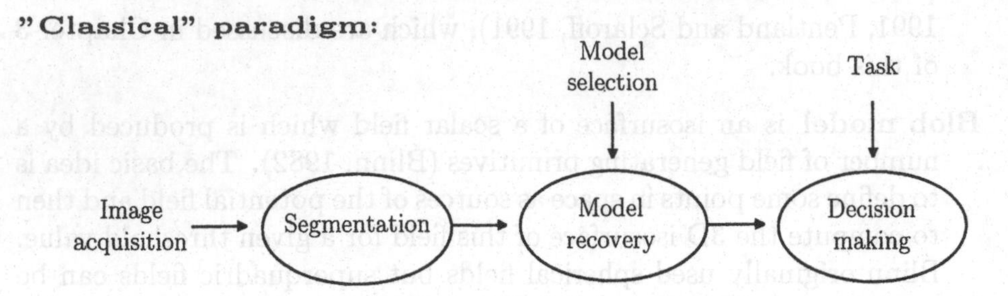

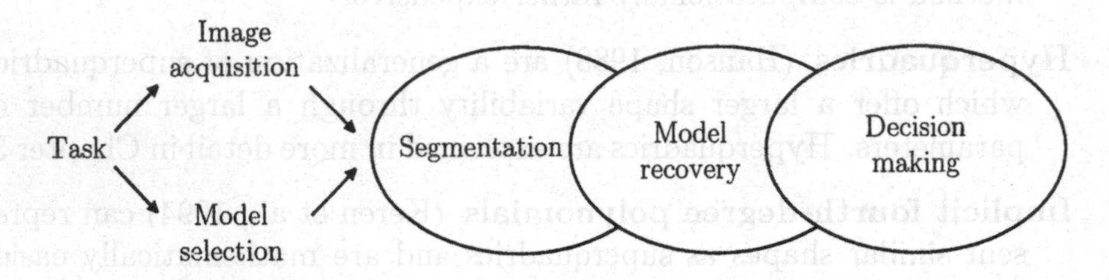

Figure 1.5. In the "classical" computer vision paradigm, segmentation, model recovery, and decision making are more or less separated. Without regard for the visual task, a nearly complete recovery of image content is striven for. In the purposive perception paradigm the model type and its granularity are determined by the task. Segmentation, model recovery and decision making are integrated.

The study of superquadric model recovery started in isolation from specific vision applications, and pre-segmented images were usually used in the initial experiments (Pentland, 1986; Boult and Gross, 1988; Solina and Bajcsy, 1990). Now, we observe that superquadric recovery is being integrated, on the one hand, with segmentation (Pentland, 1990; Gupta and Bajcsy, 1990; Leonardis et al., 1997), and, on the other hand, with decision making such as categorization (Hager, 1994), which is in accordance with ideas promoted by active and purposive vision (see Fig. 1.5).

1.3 OTHER APPLICATIONS OF SUPERQUADRICS

While in this book we concentrate on superquadric models in computer vision, one must also take note of the applications of superquadrics in computer graphics, robotics, and for modeling in general. In *graphics applications* (Barr, 1981; Pentland and Williams, 1989a; Sclaroff and Pentland, 1991; Metaxas and Terzopoulos, 1992) one of the advantages of using superquadrics is simple collision detection and dynamic simulation. Based on superquadric models, Pentland and his group built the Thing-

World system (Pentland, 1989b) which is a "real-time" physically-based solid modeler. Most commercial computer graphics software packages now offer superquadrics as one of several possible 3D building blocks.

Superquadrics are also very useful and extremely versatile primitives for constructive solid geometry modeling in *mechanical design* (Zarrugh, 1985). Because superquadric models can be directly recovered from images, they can play a useful role in *reverse engineering* applications as an intermediate step towards standard CAD models (Solina et al., 1998; Martin and Várady, 1998).

Because of their compact and intuitive parameterization, superquadric models are powerful models for building *user interfaces*, not only in modeling applications, but also for indexing 3D shape databases (Horikoshi and Kasahara, 1990). In *robotics*, superquadrics are used (sometimes in conjunction with a vision system) for grasp planning (Allen and Michelman, 1990), modeling the robot workspace (Khatib, 1986), and reasoning about function (Stark and Bowyer, 1991).

In the last chapter of the book we describe in more detail some specific applications of superquadrics.

1.4 SUMMARY

Recovering part-level representation of structures in images is highly supportive for deriving high-level symbolic description, useful for reasoning, such as object recognition, manipulation, and object avoidance. Among several part-level models that have been proposed, generalized cylinders and superquadrics are the two models most widely used in computer vision. The use of generalized cylinders is somewhat hindered by the difficulty of their recovery from images. Superquadrics, though not as powerful in representing different shapes, have some specific advantages, in particular;

- a small number of model parameters with intuitive meaning which have a large expressive power for natural shapes with rounded edges and corners, as well as for standard geometric solids with sharp edges,
- robust methods exist for reconstruction from range images, and
- superquadrics can be enhanced by adding global and local deformations.

Superquadrics had become a popular model in computer vision, especially for range image interpretation.

World system (Pentland, 1989b) which is a "real-time" physically-based solid modeler. Most commercial computer graphics software packages now offer superquadrics as one of several possible 3D building blocks.

Superquadrics are also very useful and extremely versatile primitives for constructive solid geometry modeling in *mechanical design* (Zarrugh, 1985). Because superquadric models can be directly recovered from images, they can play a useful role in *reverse engineering* applications as an intermediate step towards standard CAD models (Solina et al., 1998; Martin and Varady, 1995).

Because of their compact and intuitive parameterization, superquadric models are powerful models for building *user interfaces*, not only in modeling applications, but also for indexing 3D shape databases (Horikoshi and Kasahara, 1990). In *robotics*, superquadrics are used (sometimes in conjunction with a vision system) for grasp planning (Allen and Michelman, 1990), modeling the robot workspace (Khatib, 1986), and reasoning about function (Stark and Bowyer, 1991).

In the last chapter of the book we describe in more detail some specific applications of superquadrics.

1.4 SUMMARY

Recovering part-level representation of structures in images is highly supportive for deriving high-level symbolic description, useful for reasoning, such as object recognition, manipulation, and object avoidance. Among several part-level models that have been proposed, generalized cylinders and superquadrics are the two models most widely used in computer vision. The use of generalized cylinders is somewhat hindered by the difficulty of their recovery from images. Superquadrics, though not as powerful in representing different shapes, have some specific advantages, in particular:

- a small number of model parameters with intuitive meaning which have a large expressive power for natural shapes with rounded edges and corners, as well as for standard geometric solids with sharp edges,
- robust methods exist for reconstruction from range images, and
- superquadrics can be enhanced by adding global and local deformations.

Superquadrics had become a popular model in computer vision, especially for range image interpretation.

Chapter 2

SUPERQUADRICS AND THEIR GEOMETRIC PROPERTIES

In this chapter we define superquadrics after we outline a brief history of their development. Besides giving basic superquadric equations, we derive also some other useful geometric properties of superquadrics.

2.1 SUPERELLIPSE

A superellipse is a closed curve defined by the following simple equation

$$\left(\frac{x}{a}\right)^m + \left(\frac{y}{b}\right)^m = 1, \tag{2.1}$$

where a and b are the size (positive real number) of the major and minor axes and m is a rational number

$$m = \frac{p}{q} > 0, \quad \text{where} \begin{cases} p & \text{is an even positive integer,} \\ q & \text{is an odd positive integer.} \end{cases} \tag{2.2}$$

If $m = 2$ and $a = b$, we get the equation of a circle. For larger m, however, we gradually get more rectangular shapes, until for $m \to \infty$ the curve takes up a rectangular shape (Fig. 2.1). On the other hand, when $m \to 0$ the curve takes up the shape of a cross.

Superellipses are special cases of curves which are known in analytical geometry as Lamé curves, where m can be any rational number (Loria, 1910). Lamé curves are named after the French mathematician Gabriel Lamé, who was the first who described these curves in the early 19th century[1].

[1] *Gabriel Lamé. Examen des différentes méthodes employées pour résoudre les problémes de geometrie, Paris, 1818.*

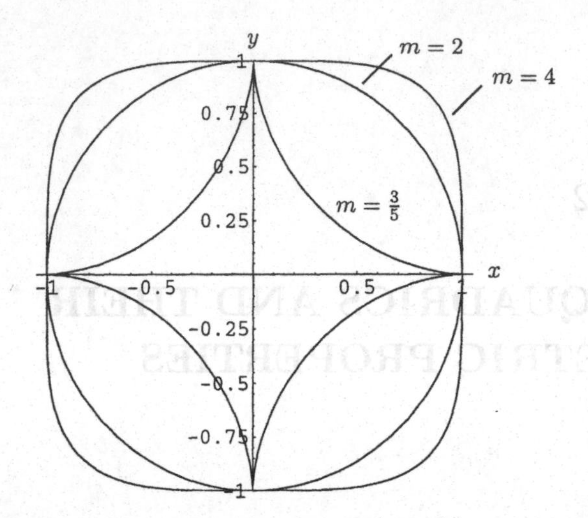

Figure 2.1. A superellipse can change continuously from a star-shape through a circle to a square shape in the limit ($m \to \infty$).

Piet Hein, a Danish scientist, writer and inventor, popularized these curves for design purposes in the 1960s (Gardner, 1965). Faced with various design problems Piet Hein proposed a shape that mediates between circular and rectangular shapes and named it a *superellipse*. Piet Hein designed the streets and an underground shopping area on Sergels Torg in Stockholm in the shape of concentric superellipses with $m = 2.5$. Other designers used superellipse shapes for design of table tops and other furniture. Piet Hein also made a generalization of superellipse to 3D which he named *superellipsoids* or *superspheres*. He named superspheres with $m = 2.5$ and the height-width ratio of 4:3 *supereggs* (Fig. 2.2). Though it looks as if a superegg standing on either of its ends should topple over, it does not because the center of gravity is lower than the center of curvature! According to Piet Hein this spooky stability of the superegg can be taken as symbolic of the superelliptical balance between the orthogonal and the round.

Superellipses were used for lofting in the preliminary design of aircraft fuselage (Flanagan and Hefner, 1967; Faux and Pratt, 1985). In 1981, Barr generalized the superellipsoids to a family of 3D shapes that he named *superquadrics* (Barr, 1981). He introduced the notation common in the computer vision literature and also used in this book. Barr saw the importance of superquadric models in particular for computer graphics and for three-dimensional design since superquadric models, which compactly represent a continuum of useful forms with rounded edges, can easily be rendered and shaded and further deformed by parametric deformations.

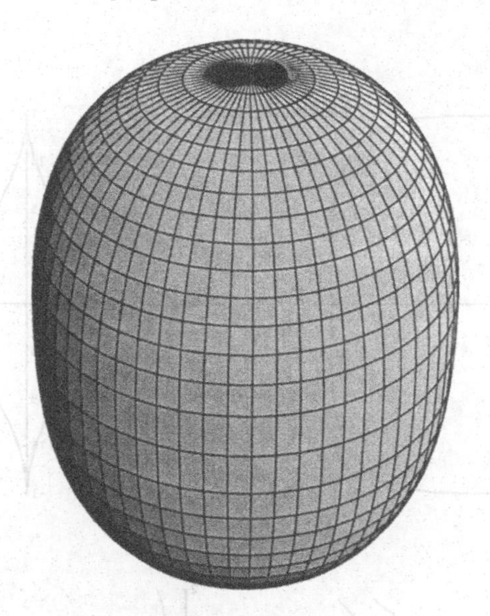

Figure 2.2. A "superegg" (superquadric with $m = 2.5$, height-width ratio = 4:3) is stable in the upright position because the center of gravity is lower than the center of curvature (Gardner, 1965).

2.1.1 LAMÉ CURVES

For Lamé curves

$$\left(\frac{x}{a}\right)^m + \left(\frac{y}{b}\right)^m = 1, \tag{2.3}$$

m can be any rational number. From the topological point of view, there are nine different types of Lamé curves depending on the form of the exponent m in equation (2.3) which is defined by positive integers $k, h \in N$ (Loria, 1910).

Lamé curves with positive m are

1. $m = \frac{2h}{2k+1} > 1$ (Fig. 2.3 a)

2. $m = \frac{2h}{2k+1} < 1$ (Fig. 2.3 b)

3. $m = \frac{2h+1}{2k} > 1$ (Fig. 2.3 c)

4. $m = \frac{2h+1}{2k} < 1$ (Fig. 2.3 d)

5. $m = \frac{2h+1}{2k+1} > 1$ (Fig. 2.3 e)

6. $m = \frac{2h+1}{2k+1} < 1$ (Fig. 2.3 f)

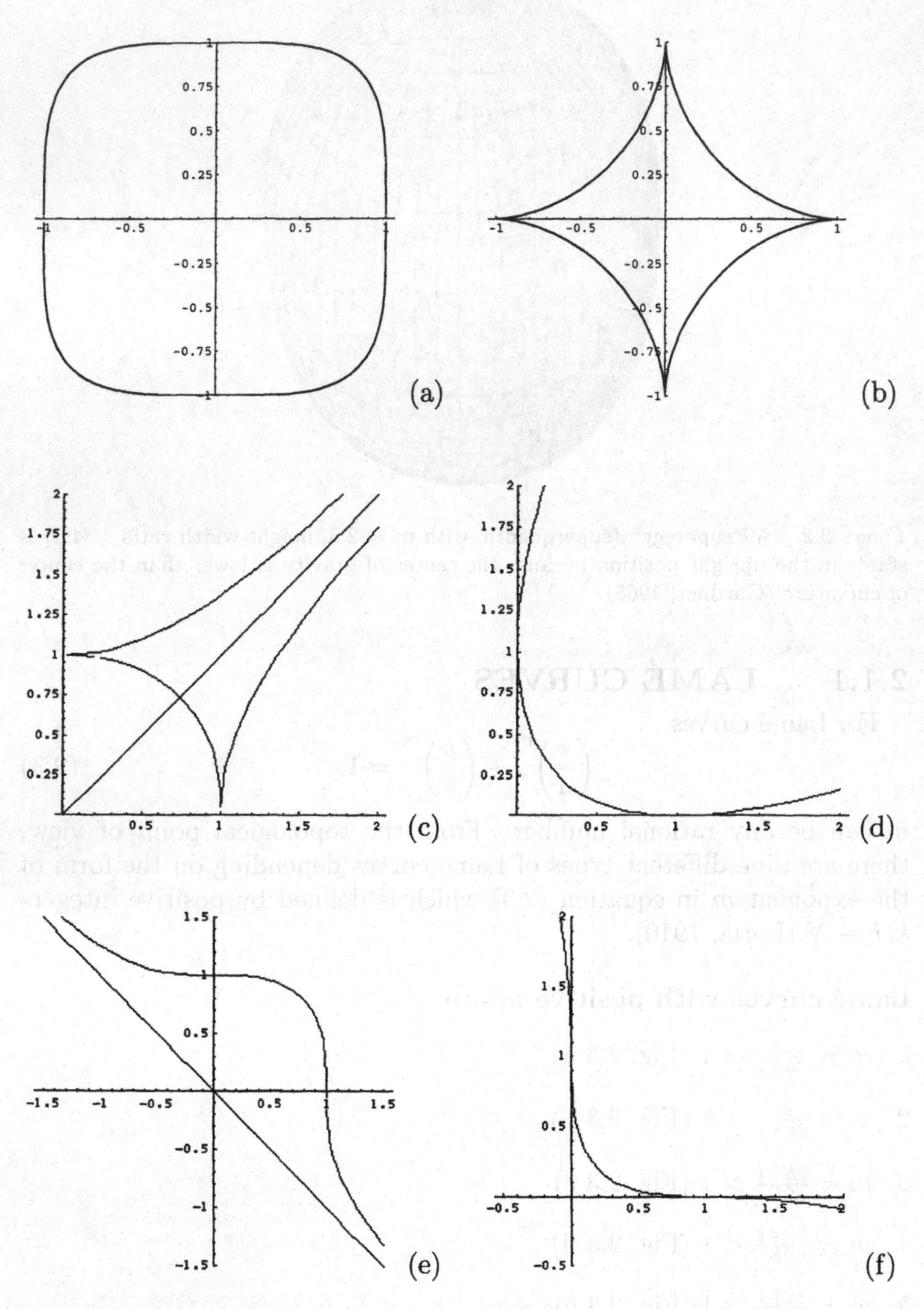

Figure 2.3. Lamé curves with positive m. Only the first two types (a) and (b) are superellipses.

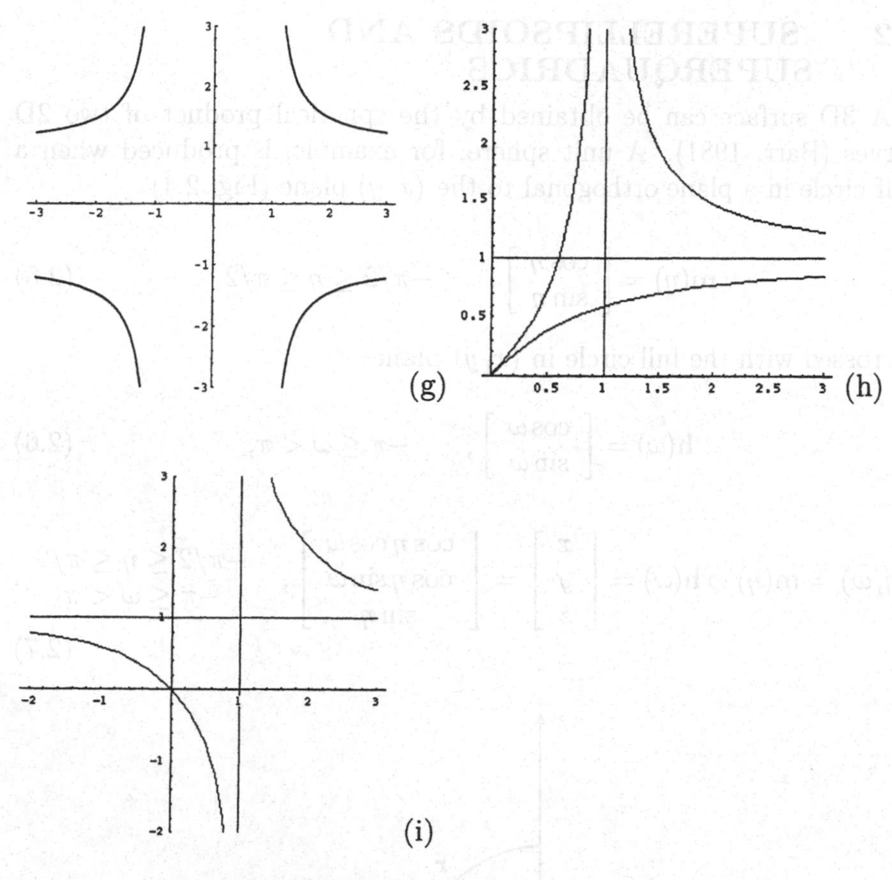

Figure 2.3 (continued). Lamé curves with negative m

Lamé curves with negative m are

7. $m = -\frac{2h}{2k+1}$ (Fig. 2.3 g)

8. $m = -\frac{2h+1}{2k}$ (Fig. 2.3 h)

9. $m = -\frac{2h+1}{2k+1}$ (Fig. 2.3 i)

which are shown in Fig. 2.3. Only the first type of Lamé curves (Fig. 2.3 a) are superellipses in the strict sense, but usually also the second type (Fig. 2.3 b) is included since the only difference is in the value of the exponent m (< 1 or > 1). Superellipses can therefore be written as

$$\left(\frac{x}{a}\right)^{\frac{2}{\varepsilon}} + \left(\frac{y}{b}\right)^{\frac{2}{\varepsilon}} = 1, \tag{2.4}$$

where ε can be any positive real number if the two terms are first raised to the second power.

2.2 SUPERELLIPSOIDS AND SUPERQUADRICS

A 3D surface can be obtained by the spherical product of two 2D curves (Barr, 1981). A unit sphere, for example, is produced when a half circle in a plane orthogonal to the (x, y) plane (Fig. 2.4)

$$\mathbf{m}(\eta) = \begin{bmatrix} \cos\eta \\ \sin\eta \end{bmatrix}, \qquad -\pi/2 \leq \eta \leq \pi/2 \tag{2.5}$$

is crossed with the full circle in (x, y) plane

$$\mathbf{h}(\omega) = \begin{bmatrix} \cos\omega \\ \sin\omega \end{bmatrix}, \qquad -\pi \leq \omega < \pi, \tag{2.6}$$

$$\mathbf{r}(\eta, \omega) = \mathbf{m}(\eta) \otimes \mathbf{h}(\omega) = \begin{bmatrix} x \\ y \\ z \end{bmatrix} = \begin{bmatrix} \cos\eta\cos\omega \\ \cos\eta\sin\omega \\ \sin\eta \end{bmatrix}, \qquad \begin{matrix} -\pi/2 \leq \eta \leq \pi/2 \\ -\pi \leq \omega < \pi \end{matrix} . \tag{2.7}$$

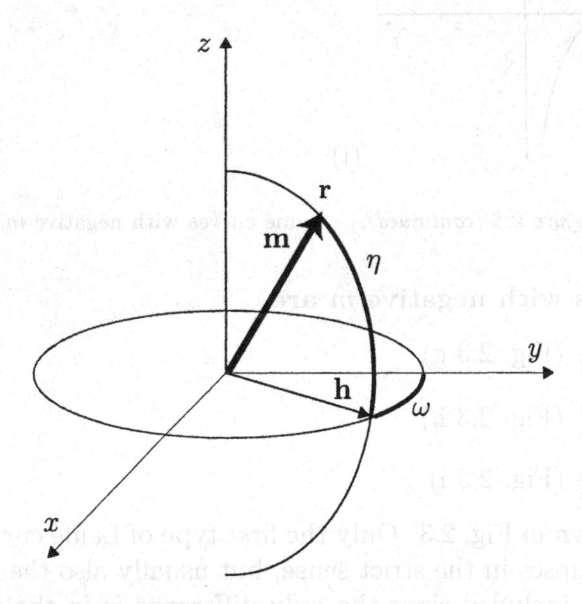

Figure 2.4. A 3D vector **r**, which defines a closed 3D surface, can be obtained by a spherical product of two 2D curves.

Analogous to a circle, a superellipse

$$\left(\frac{x}{}\right)^{\frac{2}{\varepsilon}} + \left(\frac{y}{}\right)^{\frac{2}{\varepsilon}} = 1 \tag{2.8}$$

can be written as

$$\mathbf{s}(\theta) = \begin{bmatrix} a\cos^{\varepsilon}\theta \\ b\sin^{\varepsilon}\theta \end{bmatrix}, \qquad -\pi \le \theta \le \pi \ . \tag{2.9}$$

Note that exponentiation with ε is a *signed power function* such that $\cos^{\varepsilon}\theta = \mathrm{sign}(\cos\theta)|\cos\theta|^{\varepsilon}$! Superellipsoids can therefore be obtained by a spherical product of a pair of such superellipses

$$\begin{aligned} \mathbf{r}(\eta,\omega) &= \mathbf{s}_1(\eta) \otimes \mathbf{s}_2(\omega) = \qquad (2.10) \\ &= \begin{bmatrix} \cos^{\varepsilon_1}\eta \\ a_3\sin^{\varepsilon_1}\eta \end{bmatrix} \otimes \begin{bmatrix} a_1\cos^{\varepsilon_2}\omega \\ a_2\sin^{\varepsilon_2}\omega \end{bmatrix} = \\ &= \begin{bmatrix} a_1\cos^{\varepsilon_1}\eta\cos^{\varepsilon_2}\omega \\ a_2\cos^{\varepsilon_1}\eta\sin^{\varepsilon_2}\omega \\ a_3\sin^{\varepsilon_1}\eta \end{bmatrix}, \qquad \begin{matrix} -\pi/2 \le \eta \le \pi/2 \\ -\pi \le \omega < \pi \end{matrix} \ . \end{aligned}$$

Parameters a_1, a_2 and a_3 are scaling factors along the three coordinate axes. ε_1 and ε_2 are derived from the exponents of the two original superellipses. ε_2 determines the shape of the superellipsoid cross section parallel to the (x, y) plane, while ε_1 determines the shape of the superellipsoid cross section in a plane perpendicular to the (x, y) plane and containing z axis (Fig. 2.5).

An alternative, implicit superellipsoid equation can be derived from the explicit equation using the equality $\cos^2\alpha + \sin^2\alpha = 1$. We rewrite equation (2.10) as follows:

$$\left(\frac{x}{a_1}\right)^2 = \cos^{2\varepsilon_1}\eta\,\cos^{2\varepsilon_2}\omega \ , \tag{2.11}$$

$$\left(\frac{y}{a_2}\right)^2 = \cos^{2\varepsilon_1}\eta\,\sin^{2\varepsilon_2}\omega \ , \tag{2.12}$$

$$\left(\frac{z}{a_3}\right)^2 = \sin^{2\varepsilon_1}\eta \ . \tag{2.13}$$

Raising both sides of equations (2.11) and (2.12) to the power of $1/\varepsilon_2$ and then adding respective sides of these two equations gives

$$\left(\frac{x}{a_1}\right)^{\frac{2}{\varepsilon_2}} + \left(\frac{y}{a_2}\right)^{\frac{2}{\varepsilon_2}} = \cos^{\frac{2\varepsilon_1}{\varepsilon_2}}\eta \ . \tag{2.14}$$

Next, we raise both sides of equation (2.13) to the power of $1/\varepsilon_1$ and both sides of equation (2.14) to the power of $\varepsilon_2/\varepsilon_1$. By adding the respective

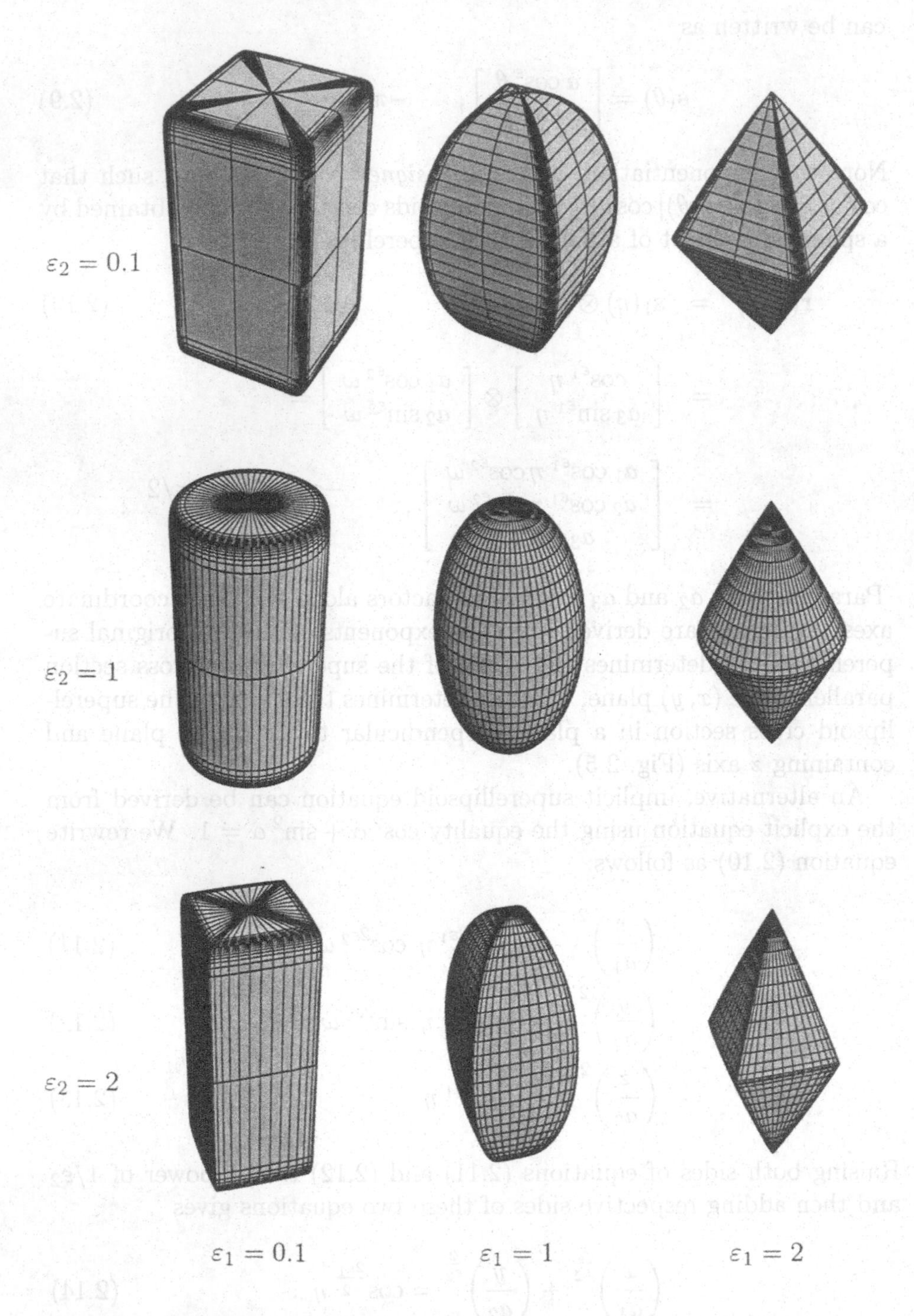

Figure 2.5. **Superellipsoids with different values of exponents ε_1 and ε_2. Size parameters a_1, a_2, a_3 are kept constant. Superquadric-centered coordinate axis z points upwards!**

sides of these two equations we get the implicit superquadric equation

$$\left(\left(\frac{x}{a_1}\right)^{\frac{2}{\varepsilon_2}} + \left(\frac{y}{a_2}\right)^{\frac{2}{\varepsilon_2}}\right)^{\frac{\varepsilon_2}{\varepsilon_1}} + \left(\frac{z}{a_3}\right)^{\frac{2}{\varepsilon_1}} = 1. \qquad (2.15)$$

All points with coordinates (x, y, z) that correspond to the above equation lie, by definition, on the surface of the superellipsoid.

The function

$$F(x, y, z) = \left(\left(\frac{x}{a_1}\right)^{\frac{2}{\varepsilon_2}} + \left(\frac{y}{a_2}\right)^{\frac{2}{\varepsilon_2}}\right)^{\frac{\varepsilon_2}{\varepsilon_1}} + \left(\frac{z}{a_3}\right)^{\frac{2}{\varepsilon_1}} \qquad (2.16)$$

is also called the *inside-outside* function because it provides a simple test whether a given point lies inside or outside the superquadric. If $F < 1$, the given point (x, y, z) is inside the superquadric, if $F = 1$ the corresponding point lies on the surface of the superquadric, and if $F > 1$ the point lies outside the superquadric.

A special case of superellipsoids when $\varepsilon_1 = \varepsilon_2$

$$\left(\frac{x}{a}\right)^{2m} + \left(\frac{y}{b}\right)^{2m} + \left(\frac{z}{c}\right)^{2m} = 1 \qquad (2.17)$$

was already studied by S. Spitzer[2]. Spitzer computed the area of superellipse and the volume of this special superellipsoid when m is a natural number (Loria, 1910).

2.2.1 SUPERQUADRICS

The term *superquadrics* was defined by Barr in his seminal paper (Barr, 1981). Superquadrics are a family of shapes that includes not only superellipsoids, but also superhyperboloids of one piece and superhyperboloids of two pieces, as well as supertoroids (Fig. 2.6). In computer vision literature, it is common to refer to superellipsoids by the more generic term of superquadrics. In this book we also use the term superquadrics as a synonym for superellipsoids.

By means of introducing parametric exponents of trigonometric functions, Barr made a generalization not only of ellipsoids, but also of the other two standard quadric surfaces; hyperboloids of one sheet

$$\left(\frac{x}{a_1}\right)^2 + \left(\frac{y}{a_2}\right)^2 - \left(\frac{z}{a_3}\right)^2 = 1 \qquad (2.18)$$

[2] *Arch. Math. Phys. LXI, 1877.*

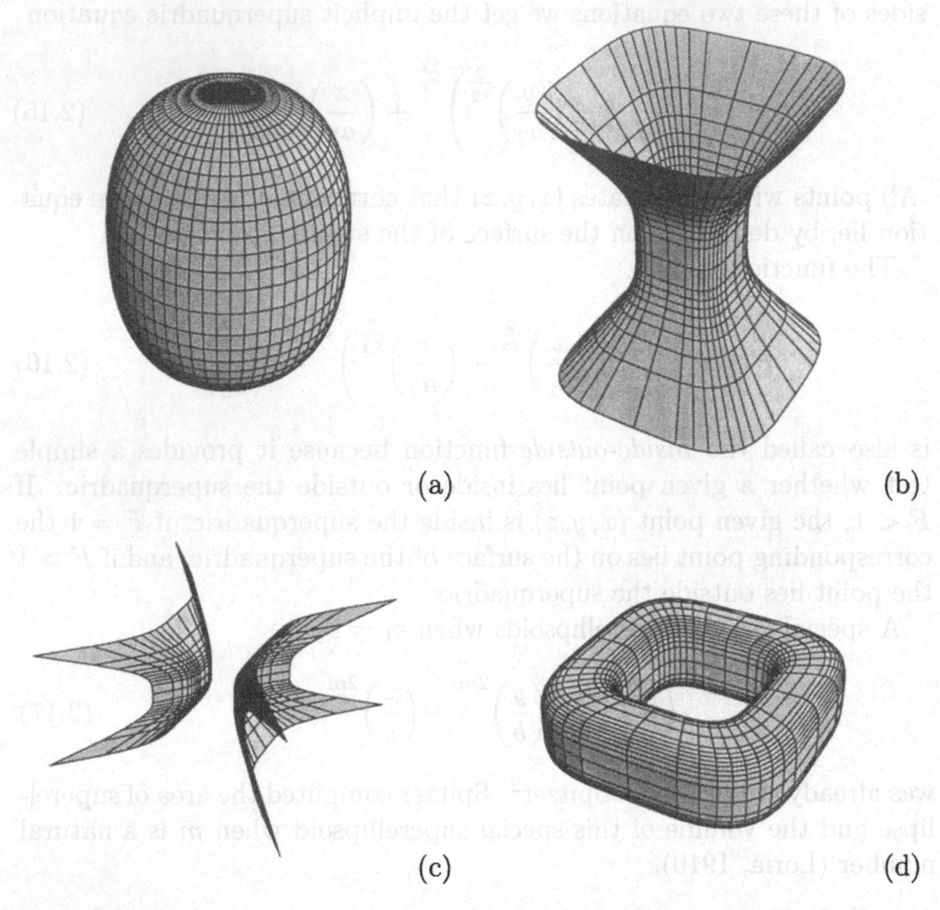

Figure 2.6. Superquadrics are a family of shapes that includes (a) superellipsoids, (b) superhyperboloids of one, and (c) of two pieces, and (d) supertoroids.

and hyperboloids of two sheets

$$\left(\frac{x}{a_1}\right)^2 - \left(\frac{y}{a_2}\right)^2 - \left(\frac{z}{a_3}\right)^2 = 1 \ . \tag{2.19}$$

Superhyperboloids of one piece are therefore defined by the surface vector

$$\begin{aligned} \mathbf{r}(\eta,\omega) &= \begin{bmatrix} \sec^{\varepsilon_1}\eta \\ a_3\tan^{\varepsilon_1}\eta \end{bmatrix} \otimes \begin{bmatrix} a_1\cos^{\varepsilon_2}\omega \\ a_2\sin^{\varepsilon_2}\omega \end{bmatrix} = \qquad (2.20) \\ &= \begin{bmatrix} a_1\sec^{\varepsilon_1}\eta\cos^{\varepsilon_2}\omega \\ a_2\sec^{\varepsilon_1}\eta\sin^{\varepsilon_2}\omega \\ a_3\tan^{\varepsilon_1}\eta \end{bmatrix}, \qquad \begin{array}{c} -\pi/2 < \eta < \pi/2 \\ -\pi \le \omega < \pi \end{array} \end{aligned}$$

and by the implicit function

$$F(x,y,z) = \left(\left(\frac{x}{a_1}\right)^{\frac{2}{\varepsilon_2}} + \left(\frac{y}{a_2}\right)^{\frac{2}{\varepsilon_2}} \right)^{\frac{\varepsilon_2}{\varepsilon_1}} - \left(\frac{z}{a_3}\right)^{\frac{2}{\varepsilon_1}} . \tag{2.21}$$

Superhyperboloids of two pieces are defined by the surface vector

$$\mathbf{r}(\eta,\omega) = \begin{bmatrix} \sec^{\varepsilon_1}\eta \\ a_3 \tan^{\varepsilon_1}\eta \end{bmatrix} \otimes \begin{bmatrix} a_1 \sec^{\varepsilon_2}\omega \\ a_2 \tan^{\varepsilon_2}\omega \end{bmatrix} = \tag{2.22}$$

$$= \begin{bmatrix} a_1 \sec^{\varepsilon_1}\eta \sec^{\varepsilon_2}\omega \\ a_2 \sec^{\varepsilon_1}\eta \tan^{\varepsilon_2}\omega \\ a_3 \tan^{\varepsilon_1}\eta \end{bmatrix}, \quad \begin{array}{ll} -\pi/2 < \eta < \pi/2 & \\ -\pi/2 < \omega < \pi/2 & \text{(sheet 1)} \\ \pi/2 < \omega < 3\pi/2 & \text{(sheet 2)} \end{array}$$

and by the implicit function

$$F(x,y,z) = \left(\left(\frac{x}{a_1}\right)^{\frac{2}{\varepsilon_2}} - \left(\frac{y}{a_2}\right)^{\frac{2}{\varepsilon_2}} \right)^{\frac{\varepsilon_2}{\varepsilon_1}} - \left(\frac{z}{a_3}\right)^{\frac{2}{\varepsilon_1}} . \tag{2.23}$$

A torus is a special case of extended quadric surface

$$(r-a)^2 = \left(\frac{z}{a_3}\right)^2 = 1 \ , \tag{2.24}$$

where

$$r = \sqrt{\left(\frac{x}{a_1}\right)^2 + \left(\frac{y}{a_2}\right)^2} \ . \tag{2.25}$$

Supertoroids are therefore defined by the following surface vector

$$\mathbf{r}(\eta,\omega) = \begin{bmatrix} a_4 + \cos^{\varepsilon_1}\eta \\ a_3 \sin^{\varepsilon_1}\eta \end{bmatrix} \otimes \begin{bmatrix} a_1 \cos^{\varepsilon_2}\omega \\ a_2 \sin^{\varepsilon_2}\omega \end{bmatrix} = \tag{2.26}$$

$$= \begin{bmatrix} a_1(a_4 + \cos^{\varepsilon_1}\eta)\cos^{\varepsilon_2}\omega \\ a_2(a_4 + \cos^{\varepsilon_1}\eta)\sin^{\varepsilon_2}\omega \\ a_3 \sin^{\varepsilon_1}\eta \end{bmatrix}, \quad \begin{array}{l} -\pi \leq \eta < \pi \\ -\pi \leq \omega < \pi \end{array}$$

and by the implicit function

$$F(x,y,z) = \left(\left(\left(\frac{x}{a_1}\right)^{\frac{2}{\varepsilon_2}} + \left(\frac{y}{a_2}\right)^{\frac{2}{\varepsilon_2}} \right)^{\frac{\varepsilon_2}{2}} - a_4 \right)^{\frac{2}{\varepsilon_1}} + \left(\frac{z}{a_3}\right)^{\frac{2}{\varepsilon_1}}, \tag{2.27}$$

where a_4 is a positive real offset value which is related to the radius of the supertoroid in the following way

$$a_4 = \frac{R}{\sqrt{a_1^2 + a_2^2}} \ . \tag{2.28}$$

2.3 SUPERQUADRICS IN GENERAL POSITION

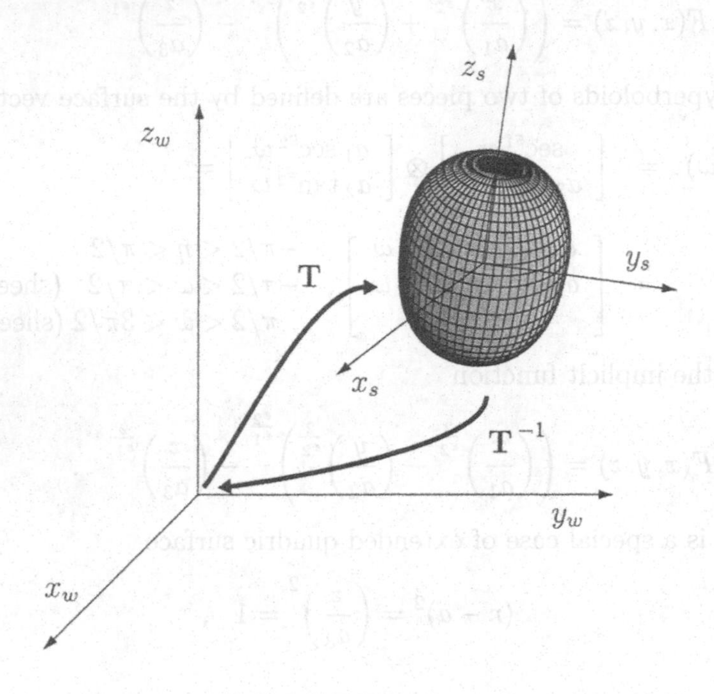

Figure 2.7. To define a superquadric in general position six additional parameters are needed.

A superellipsoid in the local or superquadric centered coordinate system (x_s, y_s, z_s) is defined by 5 parameters (3 for size in each dimension and 2 for shape defining exponents). To model or to recover superellipsoids or superquadrics from image data we must represent superquadrics in general position or in a global coordinate system. A superquadric in general position requires 6 additional parameters for expressing the rotation and translation of the superquadric relative to the center of the world coordinate system (x_w, y_w, z_w). One can use different conventions to define translation and rotation. We use a homogeneous coordinate transformation $\mathbf{T}$ to transform the 3D points expressed in the superquadric centered coordinate system $[x_s, y_s, z_s, 1]^T$ into the world coordinates $[x_w, y_w, z_w, 1]^T$ (Fig. 2.7)

$$\begin{bmatrix} x_w \\ y_w \\ z_w \\ 1 \end{bmatrix} = \mathbf{T} \begin{bmatrix} x_s \\ y_s \\ z_s \\ 1 \end{bmatrix} , \tag{2.29}$$

where

$$\mathbf{T} = \begin{bmatrix} n_x & o_x & a_x & p_x \\ n_y & o_y & a_y & p_y \\ n_z & o_z & a_z & p_z \\ 0 & 0 & 0 & 1 \end{bmatrix}. \tag{2.30}$$

For a given point, transformation $\mathbf{T}$ first rotates (defined by the parameters n, o and a) that point and then translates it for $[p_x, p_y, p_z, 1]^T$ (Paul, 1981). Since we need in our equations the points to be expressed in superquadric centered coordinates, we have to compute them from the world coordinates

$$\begin{bmatrix} x_s \\ y_s \\ z_s \\ 1 \end{bmatrix} = \mathbf{T}^{-1} \begin{bmatrix} x_w \\ y_w \\ z_w \\ 1 \end{bmatrix}. \tag{2.31}$$

Transformation $\mathbf{T}^{-1}$ performs the two operations in reverse order—it first translates a point and then rotates it.

Inverting homogeneous transformation matrix $\mathbf{T}$ gives

$$\mathbf{T}^{-1} = \begin{bmatrix} n_x & n_y & n_z & -(p_x n_x + p_y n_y + p_z n_z) \\ o_x & o_y & o_z & -(p_x o_x + p_y o_y + p_z o_z) \\ a_x & a_y & a_z & -(p_x a_x + p_y a_y + p_z a_z) \\ 0 & 0 & 0 & 1 \end{bmatrix}. \tag{2.32}$$

By substituting equations (2.29) and (2.32) into equation (2.15), we get the inside-outside function for superquadrics in general position and orientation

$$F(x_w, y_w, z_w) =$$

$$\left(\left(\frac{n_x x_w + n_y y_w + n_z z_w - p_x n_x - p_y n_y - p_z n_z}{a_1} \right)^{\frac{2}{\varepsilon_2}} + \right.$$

$$\left. + \left(\frac{o_x x_w + o_y y_w + o_z z_w - p_x o_x - p_y o_y - p_z o_z}{a_2} \right)^{\frac{2}{\varepsilon_2}} \right)^{\frac{\varepsilon_2}{\varepsilon_1}} +$$

$$+ \left(\frac{a_x x_w + a_y y_w + a_z z_w - p_x a_x - p_y a_y - p_z a_z}{a_3} \right)^{\frac{2}{\varepsilon_1}}. \tag{2.33}$$

We use Euler angles (ϕ, θ, ψ) to express the elements of the rotational part of transformation matrix $\mathbf{T}$. Euler angles define orientation in terms of rotation ϕ about the z axis, followed by a rotation θ about the new y axis, and finally, a rotation ψ about the new z axis (Paul, 1981)

$$\mathbf{T} = \begin{bmatrix} \cos\phi\cos\theta\cos\psi - \sin\phi\sin\psi & -\cos\phi\cos\theta\sin\psi - \sin\phi\cos\psi & \cos\phi\sin\theta & p_x \\ \sin\phi\cos\theta\cos\psi + \cos\phi\sin\theta & -\sin\phi\cos\theta\sin\psi + \cos\phi\cos\theta & \sin\phi\sin\theta & p_y \\ -\sin\theta\cos\psi & \sin\theta\sin\psi & \cos\theta & p_z \\ 0 & 0 & 0 & 1 \end{bmatrix}. \tag{2.34}$$

The inside-outside function for superquadrics in general position has therefore, 11 parameters

$$F(x_w, y_w, z_w) = F(x_w, y_w, z_w; a_1, a_2, a_3, \varepsilon_1, \varepsilon_2, \phi, \theta, \psi, p_x, p_y, p_z), \tag{2.35}$$

where a_1, a_2, a_3 define the superquadric size; ε_1 and ε_2 the shape; ϕ, θ, ψ the orientation, and p_x, p_y, p_z the position in space. We refer to the set of all model parameters as $\Lambda = \{\lambda_1, \lambda_2, \ldots, \lambda_{11}\}$.

2.4 SOME GEOMETRIC PROPERTIES OF SUPERELLIPSOIDS

In this section, derivations of superellipsoid normal vector, radial Euclidean distance between a point and a superellipsoid, rim of a superellipsoid in general orientation, area and inertial moments of superellipse, as well as volume and inertial moments of superellipsoids are given. Zarrugh (Zarrugh, 1985) proposed numerical methods for computing volume and moments of inertia for superellipsoids. Here we derive these properties analytically.

2.4.1 NORMAL VECTOR OF THE SUPERELLIPSOID SURFACE

Normal vector at a point $\mathbf{r}(\eta, \omega)$ on the superellipsoid surface (Eq. (2.10) on page 19) is defined by a cross product of the tangent vectors along the coordinate curves

$$\mathbf{n}(\eta, \omega) = \mathbf{r}_\eta(\eta, \omega) \times \mathbf{r}_\omega(\eta, \omega) =$$

$$= \begin{bmatrix} -a_1\varepsilon_1 \sin\eta \cos^{\varepsilon_1 - 1}\eta \cos^{\varepsilon_2}\omega \\ -a_2\varepsilon_1 \sin\eta \cos^{\varepsilon_1 - 1}\eta \sin^{\varepsilon_2}\omega \\ a_3\varepsilon_1 \sin^{\varepsilon_1 - 1}\eta \cos\eta \end{bmatrix} \times \begin{bmatrix} -a_1\varepsilon_2 \cos^{\varepsilon_1}\eta \sin\omega \cos^{\varepsilon_2 - 1}\omega \\ a_2\varepsilon_2 \cos^{\varepsilon_1}\eta \cos\omega \sin^{\varepsilon_2 - 1}\omega \\ 0 \end{bmatrix} =$$

$$= \begin{bmatrix} -a_2 a_3 \varepsilon_1 \varepsilon_2 \sin^{\varepsilon_1 - 1}\eta \cos^{\varepsilon_1 + 1}\eta \cos\omega \sin^{\varepsilon_2 - 1}\omega \\ -a_1 a_3 \varepsilon_1 \varepsilon_2 \sin^{\varepsilon_1 - 1}\eta \cos^{\varepsilon_1 + 1}\eta \sin\omega \cos^{\varepsilon_2 - 1}\omega \\ -a_1 a_2 \varepsilon_1 \varepsilon_2 \sin\eta \cos^{2\varepsilon_1 - 1}\eta \sin^{\varepsilon_2 - 1}\omega \cos^{\varepsilon_2 - 1}\omega \end{bmatrix}. \tag{2.36}$$

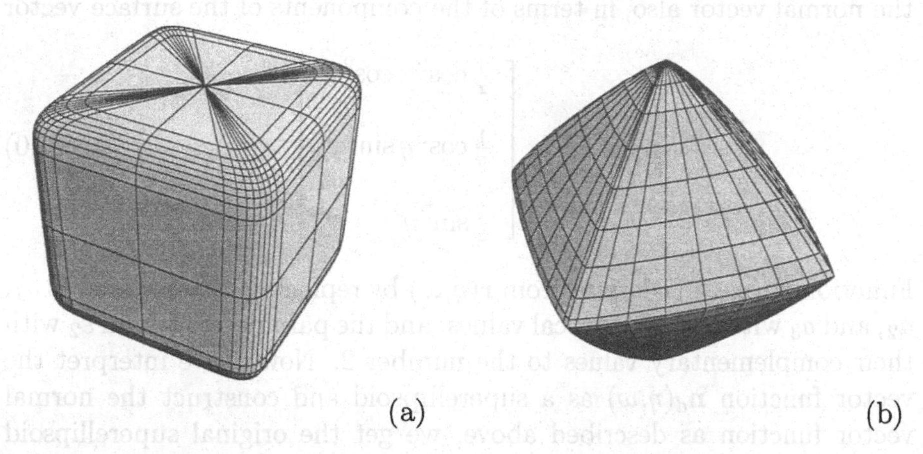

Figure 2.8. **Every (a) superquadric has a (b) dual superquadric which is defined by the scaled normal vector of the original one.**

The above expression can be simplified by defining the following common term

$$f(\eta,\omega) = -a_1 a_2 a_3 \varepsilon_1 \varepsilon_2 \sin^{\varepsilon_1 - 1}\eta \cos^{2\varepsilon_1 - 1}\eta \sin^{\varepsilon_2 - 1}\omega \cos^{\varepsilon_2 - 1}\omega\ , \qquad (2.37)$$

so that the normal vector can be written as

$$\mathbf{n}(\eta,\omega) = f(\eta,\omega) \begin{bmatrix} \frac{1}{a_1}\cos^{2-\varepsilon_1}\eta \cos^{2-\varepsilon_2}\omega \\ \frac{1}{a_2}\cos^{2-\varepsilon_1}\eta \sin^{2-\varepsilon_2}\omega \\ \frac{1}{a_3}\sin^{2-\varepsilon_1}\eta \end{bmatrix} . \qquad (2.38)$$

The scalar function $f(\eta,\omega)$ term can be dropped out if we only need the surface normal direction. By doing this, we actually get a dual superquadric to the original superquadric $\mathbf{r}(\eta,\omega)$ (Barr, 1981)

$$\mathbf{n}_d(\eta,\omega) = \begin{bmatrix} \frac{1}{a_1}\cos^{2-\varepsilon_1}\eta \cos^{2-\varepsilon_2}\omega \\ \frac{1}{a_2}\cos^{2-\varepsilon_1}\eta \sin^{2-\varepsilon_2}\omega \\ \frac{1}{a_3}\sin^{2-\varepsilon_1}\eta \end{bmatrix} . \qquad (2.39)$$

A superquadric and its dual superquadric are shown in Fig. 2.8. Using the explicit equation for superellipsoid surfaces (Eq. 2.10) we can express the normal vector also, in terms of the components of the surface vector

$$\mathbf{n}_d(\eta,\omega) = \begin{bmatrix} \frac{1}{x}\cos^2\eta\cos^2\omega \\ \\ \frac{1}{y}\cos^2\eta\sin^2\omega \\ \\ \frac{1}{z}\sin^2\eta \end{bmatrix} . \tag{2.40}$$

Function $\mathbf{n}_d(\eta,\omega)$ is derived from $\mathbf{r}(\eta,\omega)$ by replacing the parameters a_1, a_2, and a_3 with their reciprocal values, and the parameters ε_1 and ε_2 with their complementary values to the number 2. Now, if we interpret the vector function $\mathbf{n}_d(\eta,\omega)$ as a superellipsoid and construct the normal vector function as described above, we get the original superellipsoid $\mathbf{r}(\eta,\omega)$.

Note that those superquadrics which have ε_1 and $\varepsilon_2 > 2$ have sharp spiky corners where the normal vector is not uniquely defined.

2.4.2 DISTANCE BETWEEN A POINT AND A SUPERELLIPSOID

Although the true Euclidean distance between a point and a superellipsoid can be calculated by using numerical minimization, we do not know of any closed form solution in form of an algebraic expression. But for *radial Euclidean distance* such an expression can be derived based on the implicit superellipsoid equation (Whaite and Ferrie, 1991). The radial Euclidean distance is defined as a distance between a point and a superellipsoid along a line through the point and the center of a superellipsoid. We will summarize the derivation of this function and in Chapter 4 relate it to a distance measure that we proposed for recovery of superellipsoids from range data (Solina and Bajcsy, 1990).

The derivation is illustrated in Fig. 2.9. For a point defined by a vector $\mathbf{r_0} = (x_0, y_0, z_0)$ in the canonical coordinate system of a superellipsoid, we are looking for a scalar β, that scales the vector, so that the tip of the scaled vector $\mathbf{r_s} = \beta\mathbf{r_0}$ lies on the surface of the superellipsoid. Thus for the scaled vector $\mathbf{r_s}$, the following equation holds

$$F(\beta x_0, \beta y_0, \beta z_0) = \left[\left[\left(\frac{\beta x_0}{a_1}\right)^{\frac{2}{\varepsilon_2}} + \left(\frac{\beta y_0}{a_2}\right)^{\frac{2}{\varepsilon_2}} \right]^{\frac{\varepsilon_2}{\varepsilon_1}} + \left(\frac{\beta z_0}{a_3}\right)^{\frac{2}{\varepsilon_1}} \right] = 1 . \tag{2.41}$$

From this equation, it follows directly

$$F(x_0, y_0, z_0) = \beta^{-\frac{2}{\varepsilon_1}}. \tag{2.42}$$

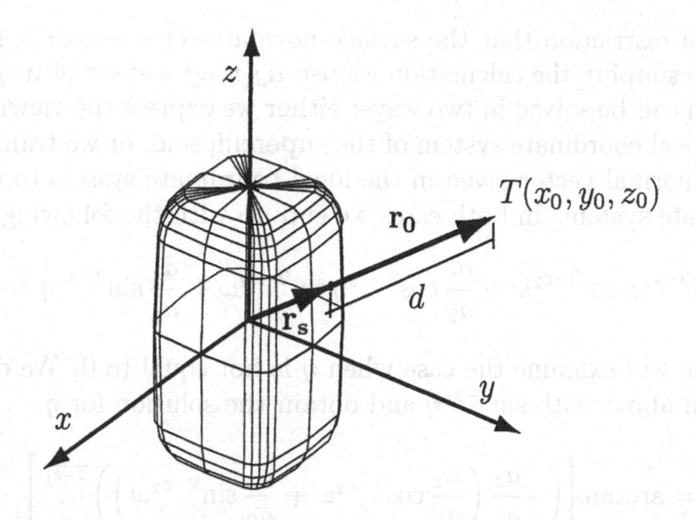

Figure 2.9. Geometric interpretation of the radial Euclidean distance

Thus the radial Euclidean distance is

$$d = |\mathbf{r_0} - \mathbf{r_S}| = |\mathbf{r_0} - \beta\mathbf{r_0}| = |\mathbf{r_0}||1 - F^{-\frac{\varepsilon_1}{2}}(x_0, y_0, z_0)| = \\ = |\mathbf{r_S}||F^{\frac{\varepsilon_1}{2}}(x_0, y_0, z_0) - 1|. \quad (2.43)$$

So for any point T in space, with given coordinates (x_0, y_0, z_0), we can determine its position relative to the superellipsoid by simply calculating the value of the $F(x_0, y_0, z_0)$. The following properties hold:

- $F(x_0, y_0, z_0) = 1 \Longleftrightarrow \beta = 1 \Longleftrightarrow$ point T belongs to the surface of the superellipsoid,
- $F(x_0, y_0, z_0) > 1 \Longleftrightarrow \beta < 1 \Longleftrightarrow$ point T is outside the superellipsoid,
- $F(x_0, y_0, z_0) < 1 \Longleftrightarrow \beta > 1 \Longleftrightarrow$ point T is inside the superellipsoid.

2.4.3 RIM OF A SUPERELLIPSOID IN GENERAL ORIENTATION

The rim is a closed space curve which partitions the object surface into a visible and invisible part. Wo would like to find the analytical form of this curve for a superellipsoid in general orientation. Assuming that we look at the superellipsoid from an infinitely distant point (orthographic projection) in the direction of the z axis of the world coordinate system. For the superellipsoid, the point is on the rim if and only if, the viewing unit vector $\mathbf{v} = (0, 0, 1)$ is perpendicular to the surface normal vector

$$\mathbf{v} \cdot \mathbf{n}(\eta, \omega) = 0, \quad (2.44)$$

with the restriction that the surface normal vector length is not equal to 0. To simplify the calculation we use $\mathbf{n}_d(\eta, \omega)$ instead of $\mathbf{n}(\eta, \omega)$. The problem can be solved in two ways: either we express the viewing vector in the local coordinate system of the superellipsoid, or we transform the surface normal vector given in the local coordinate system to the world coordinate system. In both cases we end up with the following equation

$$\frac{n_z}{a_1} \cos^{2-\varepsilon_1}\eta \cos^{2-\varepsilon_2}\omega + \frac{o_z}{a_2} \cos^{2-\varepsilon_1}\eta \sin^{2-\varepsilon_2}\omega + \frac{a_z}{a_3} \sin^{2-\varepsilon_1}\eta = 0. \quad (2.45)$$

First, we will examine the case when η is not equal to 0. We divide the equation above with $\sin^{2-\varepsilon_1}\eta$ and obtain the solution for η

$$\eta(\omega) = \arctan\left[\left(-\frac{a_3}{a_z}\left(\frac{n_z}{a_1}\cos^{2-\varepsilon_2}\omega + \frac{o_z}{a_2}\sin^{2-\varepsilon_2}\omega\right)\right)^{\frac{1}{2-\varepsilon_1}}\right] . \quad (2.46)$$

Note that the arctan function is restricted to the main branch, namely $-\frac{\pi}{2} \leq \eta \leq \frac{\pi}{2}$. If η equals 0 in the solution of equation (2.45), then

$$\frac{n_z}{a_1}\cos^{2-\varepsilon_2}\omega + \frac{o_z}{a_2}\sin^{2-\varepsilon_2}\omega = 0 \ . \quad (2.47)$$

Observing that the restricted arctan function has the value 0, if and only if the argument is equal to 0, we conclude that the equation (2.46) can be used for any $-\pi \leq \omega < \pi$ where the rim is

$$\mathbf{r}(\omega) = \mathbf{r}(\eta(\omega), \omega) \ . \quad (2.48)$$

Orthographic projection of this rim to the (xy)-plane is the occluding contour of the superellipsoid in general position.

2.4.4 AREA OF A SUPERELLIPSE

A superellipse is a parameterized planar curve, defined as

$$\begin{aligned} x &= a\cos^{\varepsilon_2}\omega \\ y &= b\sin^{\varepsilon_2}\omega \end{aligned} \qquad -\pi \leq \omega < \pi \ . \quad (2.49)$$

The easiest way of finding the area of superellipse is to use the Green formula

$$A = \tfrac{1}{2}\oint_{C'} (x\,dy - y\,dx), \quad (2.50)$$

where the integration path C' is the curve itself. To simplify the integral, we do not choose the curve C', but rather the curve C that consists of the three segments C_1, C_2 and C_3 (Fig. 2.10).

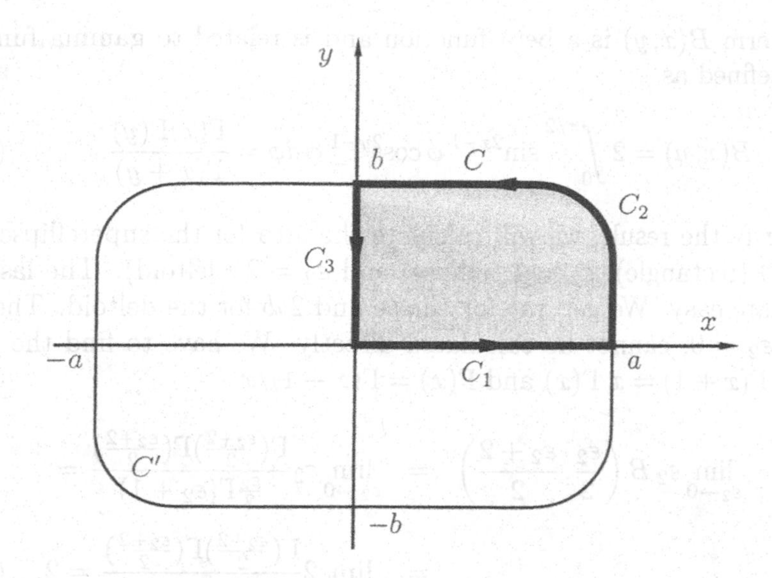

Figure 2.10. Integration path C used in Green formula to calculate the area of superellipse.

Since the superellipse is symmetric with respect to the x and y axes, it follows directly that the area equals

$$\begin{aligned} A &= 2\oint_C (x\,dy - y\,dx) = \\ &= 2\int_{C_1} (x\,dy - y\,dx) + 2\int_{C_2} (x\,dy - y\,dx) + 2\int_{C_3} (x\,dy - y\,dx). \end{aligned} \tag{2.51}$$

The integral along C_1 is equal to 0, because $y = 0$ and $dy = 0$. Similarly, the integral along C_3 is also equal to 0. So now we have

$$\begin{aligned} A &= 2\int_{C_2} (x\,dy - y\,dx) = \\ &= 2\int_0^{\pi/2} (x\dot{y} - y\dot{x})\,d\omega = \\ &= 2ab\varepsilon_2 \int_0^{\pi/2} (\sin^{\varepsilon_2 - 1}\omega \cos^{\varepsilon_2 + 1}\omega + \sin^{\varepsilon_2 + 1}\omega \cos^{\varepsilon_2 - 1}\omega)\,d\omega = \\ &= ab\varepsilon_2 \left[B\left(\frac{\varepsilon_2}{2}, \frac{\varepsilon_2 + 2}{2}\right) + B\left(\frac{\varepsilon_2 + 2}{2}, \frac{\varepsilon_2}{2}\right) \right] = \\ &= 2ab\varepsilon_2\, B\left(\frac{\varepsilon_2}{2}, \frac{\varepsilon_2 + 2}{2}\right). \end{aligned} \tag{2.52}$$

The term $B(x,y)$ is a beta function and is related to gamma function and defined as

$$B(x,y) = 2\int_0^{\pi/2} \sin^{2x-1}\phi \cos^{2y-1}\phi \, d\phi = \frac{\Gamma(x)\Gamma(y)}{\Gamma(x+y)}. \tag{2.53}$$

To verify the result, we will calculate the area for the superellipse with $\varepsilon_2 = 0$ (rectangle), $\varepsilon_2 = 1$ (ellipse) and $\varepsilon_2 = 2$ (deltoid). The last two cases are easy. We get πab for ellipse and $2ab$ for the deltoid. The first case, $\varepsilon_2 = 0$, cannot be calculated directly. We have to find the limit, using $\Gamma(x+1) = x\,\Gamma(x)$ and $\Gamma(x) = \Gamma(x+1)/x$

$$\begin{aligned}\lim_{\varepsilon_2 \to 0} \varepsilon_2 B\left(\frac{\varepsilon_2}{2}, \frac{\varepsilon_2+2}{2}\right) &= \lim_{\varepsilon_2 \to 0} \varepsilon_2 \frac{\Gamma(\frac{\varepsilon_2+2}{2})\Gamma(\frac{\varepsilon_2+2}{2})}{\frac{\varepsilon_2}{2}\Gamma(\varepsilon_2+1)} = \\ &= \lim_{\varepsilon_2 \to 0} 2\frac{\Gamma(\frac{\varepsilon_2+2}{2})\Gamma(\frac{\varepsilon_2+2}{2})}{\Gamma(\varepsilon_2+1)} = 2. \end{aligned} \tag{2.54}$$

So the area of a rectangular superellipse equals $4ab$.

2.4.5 VOLUME OF A SUPERELLIPSOID

If we cut a superellipsoid with a plane parallel to the (xy)-plane, we get a superellipse (Fig. 2.11).

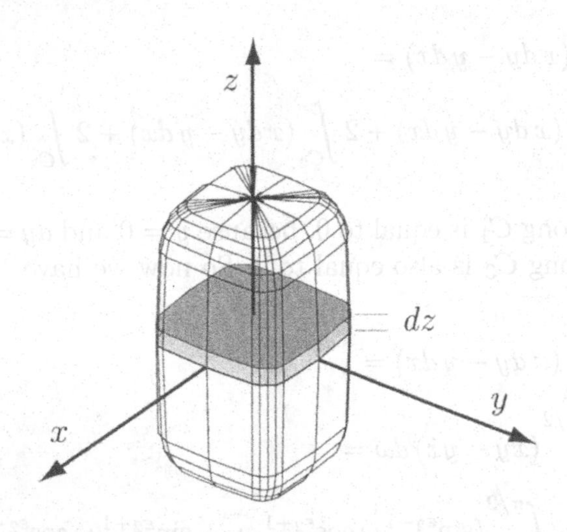

Figure 2.11. Geometric interpretation of a superellipsoid as a stack of superellipses with infinitesimal thickness dz, their size being modulated by another superellipse.

What are the parameters of this superellipse? The parameters a and b depend on the distance of the plane from the origin of the coordinate system, that is the z coordinate. The z coordinate in turn depends only on the parameter η. The area of superellipse thus equals

$$A(\eta) = 2a(\eta)b(\eta)\varepsilon_2 \, B\left(\frac{\varepsilon_2}{2}, \frac{\varepsilon_2 + 2}{2}\right), \tag{2.55}$$

where

$$a(\eta) = a_1 \cos^{\varepsilon_1} \eta, \tag{2.56}$$
$$b(\eta) = a_2 \cos^{\varepsilon_1} \eta. \tag{2.57}$$

The corresponding volume differential follows

$$\begin{aligned} dV &= A(z)\, dz = \\ &= A(\eta)\dot{z}(\eta)\, d\eta = \\ &= 2a_1a_2a_3\varepsilon_1\varepsilon_2 \, B\left(\frac{\varepsilon_2}{2}, \frac{\varepsilon_2 + 2}{2}\right) \sin^{\varepsilon_1 - 1} \eta \cos^{2\varepsilon_1 + 1} \eta \, d\eta. \end{aligned} \tag{2.58}$$

We will again use the property of superellipsoid symmetry with respect to the (xy)-plane to calculate the volume of a superellipsoid. The integration interval is from 0 to a_3 with respect to z or from 0 to $\pi/2$ with respect to η,

$$\begin{aligned} V &= 2\int_0^{a_3} A(z)\, dz = \\ &= 2\int_0^{\pi/2} A(\eta)\dot{z}(\eta)\, d\eta = \\ &= 4a_1a_2a_3\varepsilon_1\varepsilon_2 \, B\left(\frac{\varepsilon_2}{2}, \frac{\varepsilon_2+2}{2}\right)\int_0^{\pi/2} \sin^{\varepsilon_1 - 1} \eta \cos^{2\varepsilon_1+1} \eta \, d\eta = \\ &= 2a_1a_2a_3\varepsilon_1\varepsilon_2 \, B\left(\frac{\varepsilon_1}{2}, \varepsilon_1 + 1\right) B\left(\frac{\varepsilon_2}{2}, \frac{\varepsilon_2+2}{2}\right). \end{aligned} \tag{2.59}$$

By algebraic manipulation of the beta terms expressed as gamma functions, we can derive the alternative form

$$V = 2a_1a_2a_3\varepsilon_1\varepsilon_2 \, B\left(\frac{\varepsilon_1}{2} + 1, \varepsilon_1\right) B\left(\frac{\varepsilon_2}{2}, \frac{\varepsilon_2}{2}\right). \tag{2.60}$$

Fig. 2.12 shows the dependence of the ratio between the volume of a superellipsoid and a parallelepiped, with the sides $2a_1$, $2a_2$ in $2a_3$, on the shape parameters ε_1 in ε_2. Verification of the derived formula for the volume of the superellipsoids shown in Fig. 2.5 produced the values given in Table 2.1.

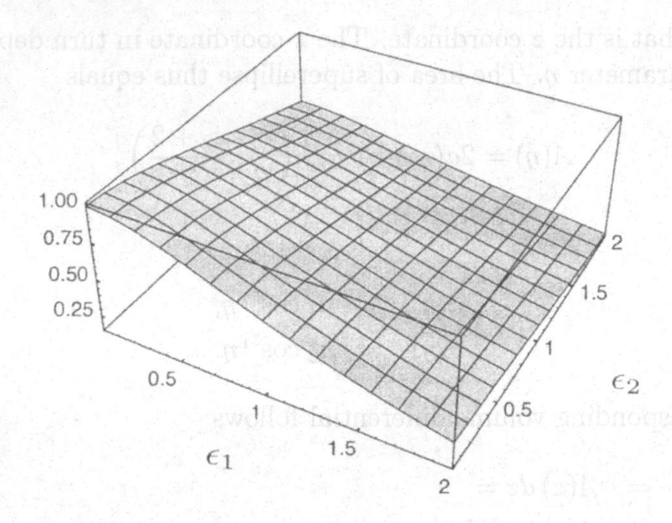

Figure 2.12. Graph of the function $\frac{1}{4}\varepsilon_1\varepsilon_2 B(\frac{\varepsilon_1}{2}, \varepsilon_1+1)B(\frac{\varepsilon_2}{2}, \frac{\varepsilon_2+2}{2})$ which shows how volume of superellipsoids depends on ε_1 and ε_2.

Table 2.1. Volumes for the family of the superquadrics shown in Fig. 2.5 and calculated by equation (2.59).

	$\varepsilon_1 = 0$	$\varepsilon_1 = 1$	$\varepsilon_1 = 2$
$\varepsilon_2 = 0$	$8a_1a_2a_3$	$\frac{16}{3}a_1a_2a_3$	$\frac{8}{3}a_1a_2a_3$
$\varepsilon_2 = 1$	$2\pi a_1a_2a_3$	$\frac{4}{3}\pi a_1a_2a_3$	$\frac{2}{3}\pi a_1a_2a_3$
$\varepsilon_2 = 2$	$4a_1a_2a_3$	$\frac{8}{3}a_1a_2a_3$	$\frac{4}{3}a_1a_2a_3$

2.4.6 MOMENTS OF INERTIA OF A SUPERELLIPSE

As we have used the expression for area of a superellipse to derive the volume of a superellipsoid, we will use expressions for moments of inertia of a superellipse to derive expressions for moments of inertia of a superellipsoid. To simplify the calculation of the moments of inertia, we introduce a "superelliptical" coordinate system similar to a circular coordinate system with coordinates r and ω instead of x and y, where the transformation between the two systems is given by

$$\begin{aligned} x &= ar\cos^{\varepsilon}\omega, \\ y &= br\sin^{\varepsilon}\omega. \end{aligned} \tag{2.61}$$

The determinant of Jacobian matrix for the transformation equals

$$|\mathbf{J}| = abr\varepsilon\sin^{\varepsilon-1}\omega\cos^{\varepsilon-1}\omega. \tag{2.62}$$

and the moments of inertia of a superellipse about the x, y and z axes are as follows

$$
\begin{aligned}
I^0_{xx} &= \int\int_S y^2 dxdy = \\
&= \int_{-\pi}^{\pi}\int_0^1 b^2 r^2 \sin^{2\varepsilon}\omega |\mathbf{J}| dr d\omega = \\
&= \frac{1}{2} ab^3 \varepsilon B\left(\frac{3\varepsilon}{2}, \frac{\varepsilon}{2}\right),
\end{aligned} \tag{2.63}
$$

$$
\begin{aligned}
I^0_{yy} &= \int\int_S x^2 dxdy = \\
&= \int_{-\pi}^{\pi}\int_0^1 a^2 r^2 \cos^{2\varepsilon}\omega |\mathbf{J}| dr d\omega = \\
&= \frac{1}{2} a^3 b \varepsilon B\left(\frac{3\varepsilon}{2}, \frac{\varepsilon}{2}\right),
\end{aligned} \tag{2.64}
$$

$$
\begin{aligned}
I^0_{zz} &= \int\int_S (x^2 + y^2) dxdy = \\
&= I^0_{xx} + I^0_{yy} = \\
&= ab(a^2 + b^2)\varepsilon B\left(\frac{3\varepsilon}{2}, \frac{\varepsilon}{2}\right).
\end{aligned} \tag{2.65}
$$

where $B(x, y)$ is a beta function. The symmetry of a superellipse with respect to the x and y axes of coordinate system causes the moment of deviation to vanish

$$I^0_{xy} = \int\int_S xy\, dxdy = 0. \tag{2.66}$$

The evaluation of the expressions (2.63), (2.64), and (2.65) for a circle, an ellipse, and a rectangle, taking limits where necessary, produces expected results listed in Table 2.2.

Table 2.2. Moments of inertia for special cases of superellipses

	Circle ($\varepsilon = 1$)	*Ellipse* ($\varepsilon = 1$)	*Rectangle* ($\varepsilon = 0$)
I^0_{xx}	$\frac{\pi}{4}r^4$	$\frac{\pi}{4}ab^3$	$\frac{4}{3}ab^3$
I^0_{yy}	$\frac{\pi}{4}r^4$	$\frac{\pi}{4}a^3b$	$\frac{4}{3}a^3b$
I^0_{zz}	$\frac{\pi}{2}r^4$	$\frac{\pi}{4}ab(a^2+b^2)$	$\frac{4}{3}ab(a^2+b^2)$

2.4.7 MOMENTS OF INERTIA OF A SUPERELLIPSOID

By slicing the superellipsoid along the z axis into slices of infinitesimal thickness dz parallel to xy plane and using Steiner's formula, moments of inertia of a superellipsoid can be determined

$$\begin{aligned}
I_{xx} &= \int\int\int_V (y^2 + z^2) dxdydz = \qquad (2.67)\\
&= \int_{-a_3}^{+a3} (\int\int_{S(z)} y^2 dxdy + \int\int_{S(z)} z^2 dxdy) dz = \\
&= \int_{-a_3}^{+a3} (I_{xx}^0(z) + z^2 A(z)) dz = \\
&= \int_{-\pi/2}^{+\pi/2} (I_{xx}^0(\eta) + z^2(\eta) A(\eta)) \dot{z}(\eta) d\eta = \\
&= \frac{1}{2} a_1 a_2 a_3 \varepsilon_1 \varepsilon_2 (a_2^2 B(\frac{3}{2}\varepsilon_2, \frac{1}{2}\varepsilon_2) B(\frac{1}{2}\varepsilon_1, 2\varepsilon_1 + 1) + \\
&\quad + 4a_3^2 B(\frac{1}{2}\varepsilon_2, \frac{1}{2}\varepsilon_2 + 1) B(\frac{3}{2}\varepsilon_1, \varepsilon_1 + 1)),
\end{aligned}$$

$$\begin{aligned}
I_{yy} &= \int\int\int_V (x^2 + z^2) dxdydz = \qquad (2.68)\\
&= \int_{-a_3}^{+a3} (\int\int_{S(z)} x^2 dxdy + \int\int_{S(z)} z^2 dxdy) dz = \\
&= \int_{-a_3}^{+a3} (I_{yy}^0(z) + z^2 A(z)) dz = \\
&= \int_{-\pi/2}^{+\pi/2} (I_{yy}^0(\eta) + z^2(\eta) A(\eta)) \dot{z}(\eta) d\eta = \\
&= \frac{1}{2} a_1 a_2 a_3 \varepsilon_1 \varepsilon_2 (a_1^2 B(\frac{3}{2}\varepsilon_2, \frac{1}{2}\varepsilon_2) B(\frac{1}{2}\varepsilon_1, 2\varepsilon_1 + 1) + \\
&\quad + 4a_3^2 B(\frac{1}{2}\varepsilon_2, \frac{1}{2}\varepsilon_2 + 1) B(\frac{3}{2}\varepsilon_1, \varepsilon_1 + 1)),
\end{aligned}$$

$$\begin{aligned}
I_{zz} &= \int\int\int_V (x^2 + y^2) dxdydz = \qquad (2.69)\\
&= \int_{-a_3}^{+a3} (\int\int_{S(z)} (x^2 + y^2) dxdy) dz = \\
&= \int_{-a_3}^{+a3} I_{zz}^0(z) dz =
\end{aligned}$$

$$= \int_{-\pi/2}^{+\pi/2} I_{zz}^0(\eta)\dot{z}(\eta)d\eta =$$

$$= \frac{1}{2}a_1a_2a_3\varepsilon_1\varepsilon_2(a_1^2+a_2^2)B(\frac{3}{2}\varepsilon_2,\frac{1}{2}\varepsilon_2)B(\frac{1}{2}\varepsilon_1, 2\varepsilon_1+1).$$

where $I_{xx}^0(\eta)$, $I_{yy}^0(\eta)$, $I_{zz}^0(\eta)$, and $A(\eta)$ are the respective moments of inertia and the area of a superellipse slice with parameters $a(\eta)$ and $b(\eta)$ given by equations (2.56) and (2.57).

The evaluation of the functions for inertial moments of a superellipsoid produced results in accordance with the well-known expressions for inertial moments of common geometric bodies like a sphere, an ellipsoid, and a cube. The results are listed in Table 2.3.

Table 2.3. Moments of inertia for special cases of superellipsoids

	Sphere	*Ellipsoid*	*Cube*
	$\varepsilon_1 = 1,\ \varepsilon_2 = 1$	$\varepsilon_1 = 1,\ \varepsilon_2 = 1$	$\varepsilon_1 = 0,\ \varepsilon_2 = 0$
I_{xx}	$\frac{8\pi}{15}r^5$	$\frac{4\pi}{15}abc(b^2+c^2)$	$\frac{1}{12}abc(b^2+c^2)$
I_{yy}	$\frac{8\pi}{15}r^5$	$\frac{4\pi}{15}abc(a^2+c^2)$	$\frac{1}{12}abc(a^2+c^2)$
I_{zz}	$\frac{8\pi}{15}r^5$	$\frac{4\pi}{15}abc(a^2+b^2)$	$\frac{1}{12}abc(a^2+b^2)$

2.5 COMPUTATION AND RENDERING OF SUPERQUADRICS

In any implementation using superquadrics one must be careful about the numerical evaluation of superquadric equations. All exponential terms in the implicit superquadric equations are of the form x^{2r}, where r can be any positive real number. In numerical computations one must take care of the correct order of evaluation of these exponential terms and compute them as $(x^2)^r$ to assure that the result is not a complex number but a real one when $x < 0$!

In the explicit superquadric equations one should assume that any exponentiation represents in fact, a *signed power function*

$$x^p = \text{sign}(x)|x|^p = \begin{cases} |x|^p & x \geq 0 \\ -|x|^p & x < 0 \end{cases}.$$

This detail is often missing in articles related to superquadrics, but is crucial for any software implementation.

For applications in computer vision, the values for ε_1 and ε_2 are normally bounded: $0 < \{\varepsilon_1, \varepsilon_2\} < 2$, so that only convex shapes are produced (Fig. 2.5). To prevent numerical overflow and difficulties with singularities, ε_1 and ε_2 are often further bounded $(0.1 < \{\varepsilon_1, \varepsilon_2\} < 1.9)$.

Wire-frame models are normally used for rendering superquadrics in computer vision. Rendering accuracy can be controlled simply by changing the sampling rate of the chosen parameterization. Hidden surfaces can be removed by checking the normal vectors which are easy to compute since they are dual to the surface vector (Eq. 2.39). Using the surface normal vector one can also easily generate shaded superquadric models (Fig. 2.13).

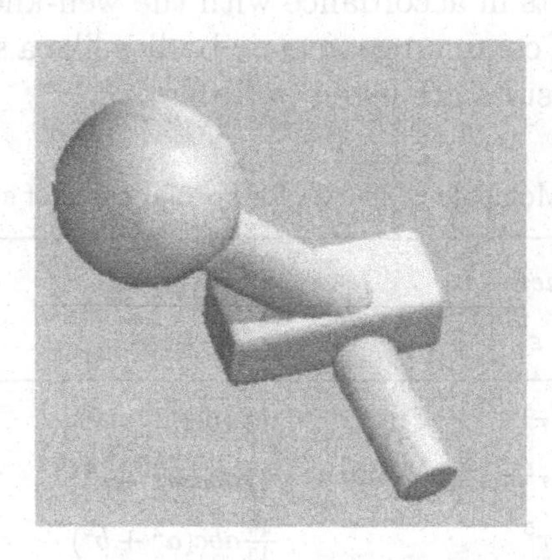

Figure 2.13. A shaded superquadric model using the Phong illumination model and three point sources of illumination.

If the angle parameters η and ω in the explicit equation (2.10) are uniformly sampled, we obtain a "trigonometric" parameterization. With trigonometric parameterization the lines of the wire-frame models are closer together in the more curved regions. This is a good feature for wire-frame models since it gives a good indication of the curvature of the model's surface. For some computer graphics applications such parameterization does not always produce satisfying results. Texturing of superquadric shapes, for example, requires a more uniform parameterization density of the surface. Other types of parameterization of superquadric surfaces were proposed which resulted in an almost uniform parameterization (Franklin and Barr, 1981; Löffelmann and Gröller, 1994; Montiel et al., 1998).

A uniform and dense parameterization of a superquadric surface can be obtained if the explicit equation, where z is as a function of x and y,

$$z = a_3 \left(1 - \left(\left(\frac{x}{a_1} \right)^{\frac{2}{\varepsilon_2}} + \left(\frac{y}{a_2} \right)^{\frac{2}{\varepsilon_2}} \right)^{\frac{\varepsilon_2}{\varepsilon_1}} \right)^{\frac{\varepsilon_1}{2}} \tag{2.70}$$

is expressed as a binomial expansion (Franklin and Barr, 1981). Since only up to the first five coefficients of the expansion need to be computed for a high resolution display, one can easily evaluate equation (2.70) for every pixel (x, y).

The "Angle, Center" parameterization (Löffelmann and Gröller, 1994) uses the angles η and ω of a ray r through the point $\mathbf{r}(\eta, \omega)$ on the superquadric surface. For a superellipse, for example, a parameterization point is defined as

$$\mathbf{r}(\omega) = r(\omega) \begin{bmatrix} \cos\omega \\ \sin\omega \end{bmatrix}, \tag{2.71}$$

where

$$r(\omega) = \frac{1}{\sqrt{\left(\left(\frac{\cos\omega}{a_1}\right)^{\frac{2}{\varepsilon_2}} + \left(\frac{\sin\omega}{a_2}\right)^{\frac{2}{\varepsilon_2}} \right)^{\varepsilon_2}}}. \tag{2.72}$$

Superquadrics can be parameterized in the same way.

A regular distribution of parameters along the superquadric surface can be obtained by treating superquadrics as a deformation of ellipsoids (Montiel et al., 1998). This linear-arc length parameterization has also a lower computational cost since the evaluation of rational exponents is avoided. An equal-distance sampling of superellipse models was also proposed by Pilu and Fisher (Pilu and Fisher, 1995).

Appendix A contains the source code for display of superquadric models in the *Mathematica* software package.

2.6 SUMMARY

In computer vision, shape models are chosen according to their degree of uniqueness and compact representation, their local support, expressiveness, and preservation of information. Superquadrics are an extension of quadric surfaces that can model a large variety of generic shapes which are useful for volumetric part representation of natural and man-made objects.

Superquadrics are defined by the explicit or implicit equation. The implicit form (Eq. 2.15) is important for the recovery of superquadrics and testing for intersection, while the explicit form (Eq. 2.10) is more suitable for rendering. We derived geometric properties such as area and moments of a superellipse and radial Euclidean distance from a point to a superellipsoid, superellipsoid volume and superellipsoid moments of inertia. These properties are useful not only for the recovery of superquadrics, but also for other tasks such as range image registration and object recognition.

is expressed as a binomial expansion (Franklin and Barr, 1981). Since only up to the first five coefficients of the expansion need to be computed for a high resolution display, one can easily evaluate equation (2.70) for every pixel (x, y).

The "Angle Center" parameterization (Löffelmann and Gröller, 1994) uses the angles η and ω of a ray r through the point $r(\eta, \omega)$ on the superquadric surface. For a superellipse, for example, a parameterization point is defined as

$$\mathbf{r}(\omega) = r(\omega) \begin{bmatrix} \cos\omega \\ \sin\omega \end{bmatrix} \tag{2.71}$$

where

$$r(\omega) = \frac{1}{\sqrt{\left(\left(\frac{\cos\omega}{a_1}\right)^{\frac{2}{\epsilon_2}} + \left(\frac{\sin\omega}{a_2}\right)^{\frac{2}{\epsilon_2}}\right)^{\epsilon_2}}} \tag{2.72}$$

Superquadrics can be parameterized in the same way.

A regular distribution of parameters along the superquadric surface can be obtained by treating superquadrics as a deformation of ellipsoids (Montiel et al., 1998). This linear-arc length parameterization has also a lower computational cost since the evaluation of rational exponents is avoided. An equal-distance sampling of superellipse models was also proposed by Pilu and Fisher (Pilu and Fisher, 1995).

Appendix A contains the source code for display of superquadric models in the *Mathematica* software package.

2.6 SUMMARY

In computer vision, shape models are chosen according to their degree of uniqueness and compact representation, their local support, expressiveness, and preservation of information. Superquadrics are an extension of quadric surfaces that can model a large variety of generic shapes which are useful for volumetric part representation of natural and man-made objects.

Superquadrics are defined by the explicit or implicit equation. The implicit form (Eq. 2.15) is important for the recovery of superquadrics and testing for intersection, while the explicit form (Eq. 2.10) is more suitable for rendering. We derived geometric properties such as area and moments of a superellipse and radial Euclidean distance from a point to a superellipsoid, superellipsoid volume and superellipsoid moments of inertia. These properties are useful not only for the recovery of superquadrics, but also for other tasks such as range image registration and object recognition.

Chapter 3

EXTENSIONS OF SUPERQUADRICS

The modeling capabilities of superquadrics can be enhanced with global and local deformations. Global deformations such as tapering, bending or twisting require just a few additional parameters in the superquadric equations. Local deformation, in general, overlay the original superquadric with a new parameterization grid which enables local change of shape. Therefore, local deformations are by its nature not tightly integrated with superquadrics and are in this book just discussed at the end of this chapter. Hyperquadrics, which include superquadrics as a special case, and are generated by taking hyperslices of high-dimensional algebraic hypersurfaces, are also described at the end of the chapter, as well as ratioquadrics which are very similar to superquadrics but have continuous first derivatives everywhere on the surface.

3.1 GLOBAL DEFORMATIONS

Global deformations require, in general, just a few parameters which are applied directly to superquadric surfaces (Barr, 1984). Of interest are mainly deformations such as simplified tapering, bending, and twisting that occur often in nature or are used for manufacturing of man-made objects.

A shape deformation is a function $\mathbf{D}$ which explicitly modifies the global coordinates of all surface points in space

$$\mathbf{X} = \mathbf{D}(\mathbf{x}) = \begin{bmatrix} X(x,y,z) \\ Y(x,y,z) \\ Z(x,y,z) \end{bmatrix}, \tag{3.1}$$

where $\mathbf{x}$ are the points (x, y, z) on the surface of the non-deformed solid and $\mathbf{X}$ are the corresponding points (X, Y, Z) after the deformation.

Both $\mathbf{x}$ and $\mathbf{X}$ are expressed in the object centered coordinate system. Any translation or rotation of the superquadric must be performed after the deformation. This can be described schematically (Witkin et al., 1987) as

$$\text{Trans}(\text{Rot}(\text{Deform}(\mathbf{x}))). \tag{3.2}$$

Tangent and normal vectors at every point on the deformed surface, which are important for rendering or for checking the consistency of the model with the input range points, can be computed from the tangent and normal vectors of the non-deformed model merely by a matrix multiplication. The surface tangent vector $\mathbf{T}(\eta, \omega)$ on the deformed superquadric can be computed by the chain rule for derivation

$$\mathbf{T}(\eta, \omega) = \frac{\partial \mathbf{X}}{\partial \eta} = \frac{\partial \mathbf{X}}{\partial \mathbf{x}} \frac{\partial \mathbf{x}}{\partial \eta} = \mathbf{J} \frac{\partial \mathbf{x}}{\partial \eta} \tag{3.3}$$

where $\mathbf{J}$ denotes the Jacobian matrix of the deformation function $\mathbf{D}$. The Jacobian matrix $\mathbf{J}$ for the transformation function $\mathbf{X} = \mathbf{D}(\mathbf{x})$ is a function of $\mathbf{x}$, and is calculated by taking partial derivatives of $\mathbf{D}$ with respect to the coordinates x, y, and z,

$$\mathbf{J}(\mathbf{x}) = \begin{bmatrix} \frac{\partial X}{\partial x} & \frac{\partial X}{\partial y} & \frac{\partial X}{\partial z} \\ \frac{\partial Y}{\partial x} & \frac{\partial Y}{\partial y} & \frac{\partial Y}{\partial z} \\ \frac{\partial Z}{\partial x} & \frac{\partial Z}{\partial y} & \frac{\partial Z}{\partial z} \end{bmatrix}. \tag{3.4}$$

The surface normal vector $\mathbf{N}$ of the deformed superquadric is computed by taking the cross product of two surface tangent vectors

$$\mathbf{N}(\eta, \omega) = \frac{\partial \mathbf{X}}{\partial \eta} \times \frac{\partial \mathbf{X}}{\partial \omega} \tag{3.5}$$

The surface normal vector $\mathbf{N}$ of the deformed superquadric can be computed also from the surface normal vector $\mathbf{n}$ of the original superquadric by multiplying it with the transpose of the inverse Jacobian (Barr, 1984)

$$\mathbf{N}(\eta, \omega) = \det \mathbf{J} \ \mathbf{J}^{-1T} \ \mathbf{n}(\eta, \omega). \tag{3.6}$$

When Barr defined his global deformations (Barr, 1984), he had in mind computer graphics applications. In computer graphics the designer has the freedom to place and orient the primitives to his liking. During shape recovery such interaction is not possible. Barr, for example, defined bending in a single plane. However, during iterative shape recovery from data points using a minimization method, the initial model is set into the world coordinate system and then incrementally adapted to the data points. During such incremental adaptation, the superquadric model cannot rotate freely in order to align the plane where bending is defined with the plane where bending is required by the data.

When deformation parameters must be recovered directly from the input data through shape model recovery, then global deformations must be defined so that the object centered coordinate system constrains as little as possible the execution of deformations. In other words, the solution space around acceptable solutions should be convex so that the model parameters can freely adapt to the actual shape. For example, we introduced an additional parameter for the bending deformation, which allows for bending in any plane that goes through the z axis of the object centered coordinate system. Tapering deformation was enhanced to enable different tapering in x and y directions so that modeling of wedge shapes is possible.

3.1.1 TAPERING

Tapering deformation gradually thins or expands an object along a chosen dimension. A tapered superquadric obtained by our tapering deformation is shown in Fig. 3.1.

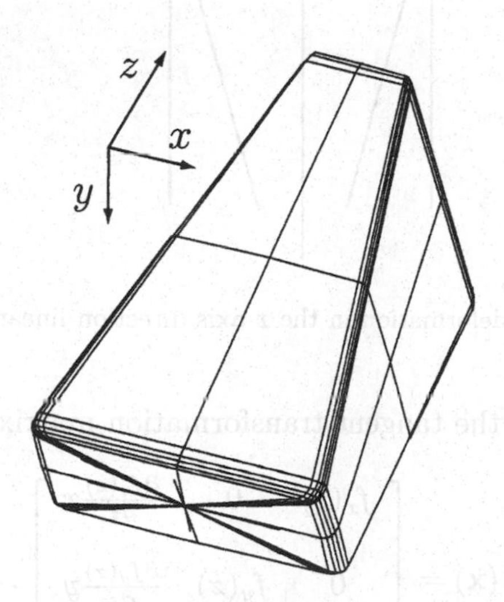

Figure 3.1. A tapered superquadric model with different tapering functions along the x and y axes.

We define the tapering deformation as a function of z

$$\begin{aligned} X &= f_x(z)\, x \\ Y &= f_y(z)\, y \\ Z &= \quad z, \end{aligned} \tag{3.7}$$

where x, y, z are the components of the original surface vector $\mathbf{x}$ and X, Y, Z are the components of the surface vector $\mathbf{X}$ of the deformed superquadric. f_x and f_y are the tapering functions in the x_S and y_S axis direction of the object-centered coordinate system (Fig. 3.2).

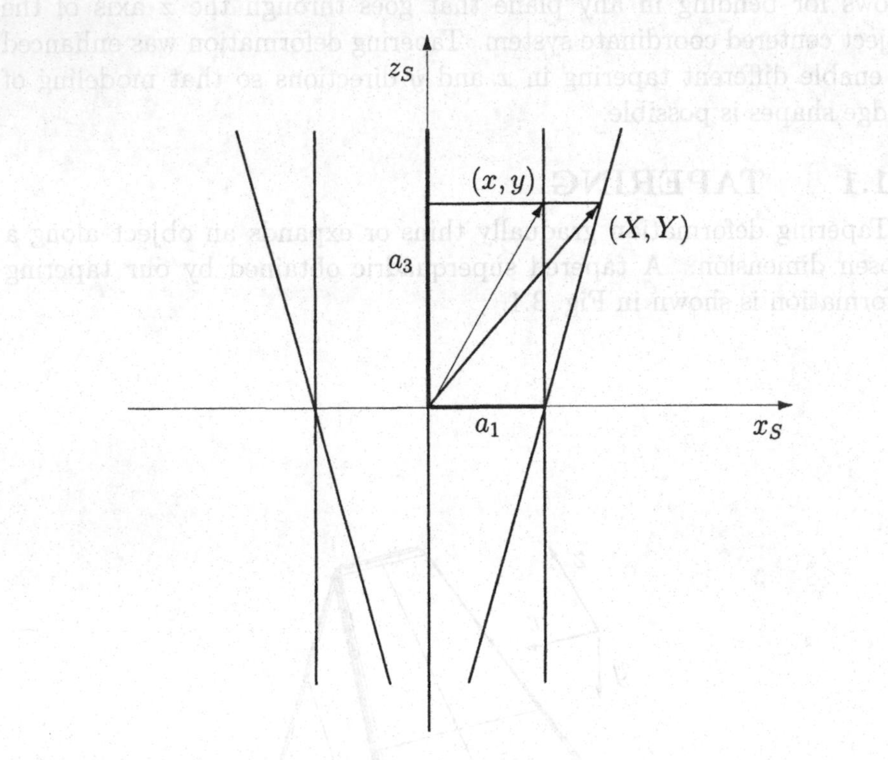

Figure 3.2. Tapering deformation in the x axis direction linearly offsets points as a function of z.

The Jacobian or the tangent transformation matrix is given by

$$\mathbf{J}(\mathbf{x}) = \begin{bmatrix} f_x(z) & 0 & \frac{\partial f_x(z)}{\partial z} x \\ 0 & f_y(z) & \frac{\partial f_y(z)}{\partial z} y \\ 0 & 0 & 1 \end{bmatrix}. \tag{3.8}$$

To be able to recover the deformation parameters, the original surface vector components x, y, z must be expressed in terms of the deformation parameters f_x, f_z and coordinates of points on the surface of a deformed superquadric X, Y, Z. The inverse transformation is given by

$$\begin{aligned} x &= \frac{X}{f_x(z)} \\ y &= \frac{Y}{f_y(z)} \\ z &= Z. \end{aligned} \tag{3.9}$$

For linear tapering, the two tapering functions are

$$f_x(z) = \frac{K_x}{a_3} z + 1 \tag{3.10}$$

$$f_y(z) = \frac{K_y}{a_3} z + 1, \tag{3.11}$$

where $-1 \leq K_x, K_y \leq 1$. When expressions for x, y, z are inserted into equation (2.35), we get the inside-outside function for a tapered superquadric in general position

$$F = F(X, Y, Z; \lambda_1, \ldots, \lambda_{11}, K_x, K_y). \tag{3.12}$$

3.1.2 BENDING

For bending, we use a deformation which transforms a straight line (the z axis of the original superquadric) into a circular section (Fig. 3.3). The curvature of the circular section is defined by k. The length of the circular superquadric spine remains the same as the previous straight spine along the z axis (a_3). Beware that such simplified bending does not correspond exactly to shape deformations when real physical objects are bent.

The bending plane rotates around coordinate axis z_S and its direction is defined by angle α (Fig. 3.4). The bending deformation is performed first, by projecting the x and y components of all points onto the bending plane, performing the bending deformation in that plane, and then projecting the points back to the original plane. The projection of a point (x, y) on the bending plane is

$$r = \sqrt{x^2 + y^2} \cos(\alpha - \beta), \tag{3.13}$$

where

$$\beta = \arctan \frac{y}{x}. \tag{3.14}$$

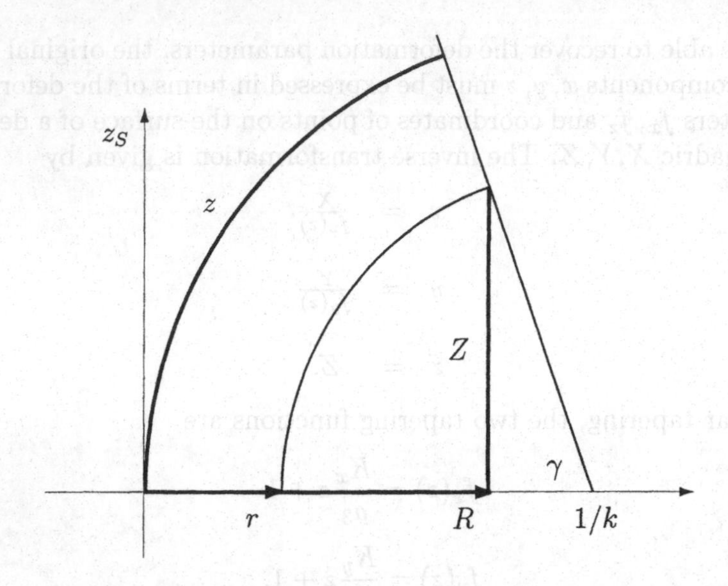

Figure 3.3. In the bending plane the superquadric spine along the z axis is bent into a circular section of the same length. Other circular sections which are offset from the original z axis expand or contract correspondingly.

Bending transforms r into

$$R = \frac{1}{k} - (\frac{1}{k} - r)\cos(\gamma), \tag{3.15}$$

where γ is the bending angle, computed from the curvature parameter k (Fig. 3.3),

$$\gamma = \frac{z}{k}. \tag{3.16}$$

By projecting R back onto the original plane, which is parallel to the bending plane, we get the transformed surface vector

$$\begin{aligned} X &= x + (R - r)\cos(\alpha) \\ Y &= y + (R - r)\sin(\alpha) \\ Z &= (\tfrac{1}{k} - r)\sin(\gamma). \end{aligned} \tag{3.17}$$

Three superquadrics bent in three different bending planes are shown in Fig. 3.5.

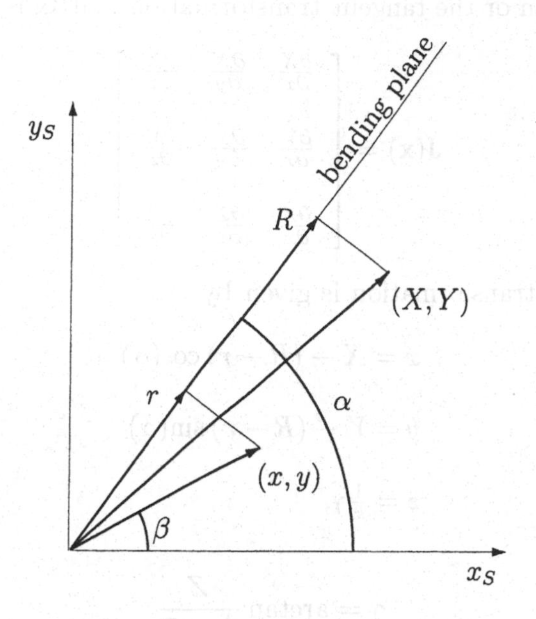

Figure 3.4. A point on the original superquadric is projected first onto the bending plane: $(x, y) \longrightarrow r$; then bending is performed: $r \longrightarrow R$; finally, the new point is projected back to the original plane of the point which is parallel to the bending plane: $R \longrightarrow (X, Y)$.

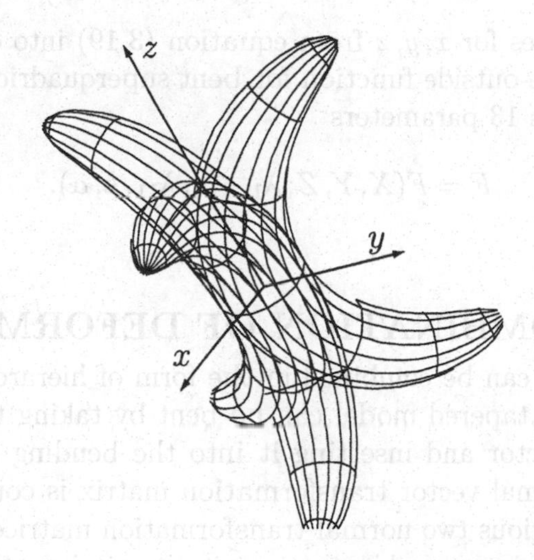

Figure 3.5. A superquadric model which is bent in three different bending planes. The bending plane rotates freely around z axis of the superquadric centered coordinate system.

The Jacobian or the tangent transformation matrix is given by

$$\mathbf{J}(\mathbf{x}) = \begin{bmatrix} \frac{\partial X}{\partial x} & \frac{\partial X}{\partial y} & \frac{\partial X}{\partial z} \\ \frac{\partial Y}{\partial x} & \frac{\partial Y}{\partial y} & \frac{\partial Y}{\partial z} \\ \frac{\partial Z}{\partial x} & \frac{\partial Z}{\partial y} & \frac{\partial Z}{\partial z} \end{bmatrix}. \tag{3.18}$$

The inverse transformation is given by

$$x = X - (R - r)\cos(\alpha)$$
$$y = Y - (R - r)\sin(\alpha) \tag{3.19}$$
$$z = \tfrac{1}{k}\gamma,$$

where

$$\gamma = \arctan \frac{Z}{\frac{1}{k} - R} \tag{3.20}$$

$$r = \frac{1}{k} - \sqrt{Z^2 + (\frac{1}{k} - R)^2} \tag{3.21}$$

$$R = \sqrt{X^2 + Y^2} \cos\left(\alpha - \arctan \frac{Y}{X}\right). \tag{3.22}$$

Inserting values for x, y, z from equation (3.19) into equation (2.35), we get the inside-outside function for bent superquadrics in general position, which has 13 parameters

$$F = F(X, Y, Z; \lambda_1, \ldots, \lambda_{11}, k, \alpha). \tag{3.23}$$

3.1.3 COMBINATION OF DEFORMATIONS

Deformations can be combined in the form of hierarchical structures (Barr, 1984). A tapered model can be bent by taking the tapered surface position vector and inserting it into the bending equations. The new surface normal vector transformation matrix is computed by multiplying the previous two normal transformation matrices. Since matrix multiplication is in general not commutative, it is not surprising that the composition of deformations is not commutative (Fig. 3.6)

$$\text{Bend}(\text{Taper}(\mathbf{x})) \neq \text{Taper}(\text{Bend}(\mathbf{x})). \tag{3.24}$$

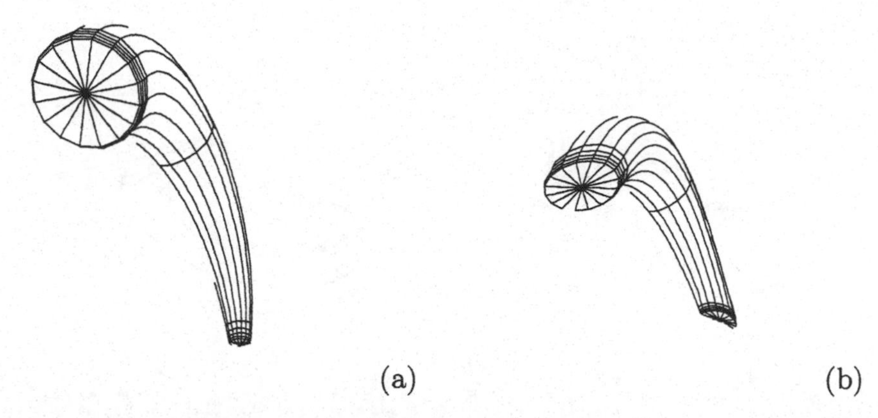

Figure 3.6. Composition of shape deformations is not commutative. When a superquadric (a) is subjected first to tapering and next to bending, the resulting shape is not the same as if the same superquadric (b) is subjected to the same deformations but in the reverse order.

Therefore, we should settle for a specific ordering of deformations. Tapering usually precedes bending, or in other words, tapering is perceived as to affect the model along the longest axis of symmetry for elongated objects, whether the axis is bent or not. Leyton (Leyton, 1988) has shown that deformations acting on prototypical shapes have a specific order—such that the transformation is more structure-preserving. We have selected the following structure for deformations

$$\mathrm{Trans}(\mathrm{Rot}(\mathrm{Bend}(\mathrm{Taper}(\mathbf{x})))). \tag{3.25}$$

The inside-outside function for superquadrics in general position which includes tapering and bending deformation has a total of 15 parameters

$$F = F(X, Y, Z; \lambda_1, \ldots, \lambda_{11}, K_x, K_y, k, \alpha). \tag{3.26}$$

3.2 LOCAL DEFORMATIONS

Addition of global deformations on the basic superquadric is sufficient only for capturing the global shape. Any local details are smoothed out by the global model (Fig. 3.7).

While a certain coarseness of models is desirable for segmentation or for object classification which employs part-level models, most object recognition problems require a much finer representation of surface details, besides other features, such as texture and color. This can be achieved by "wrapping" the superquadrics with an additional layer of

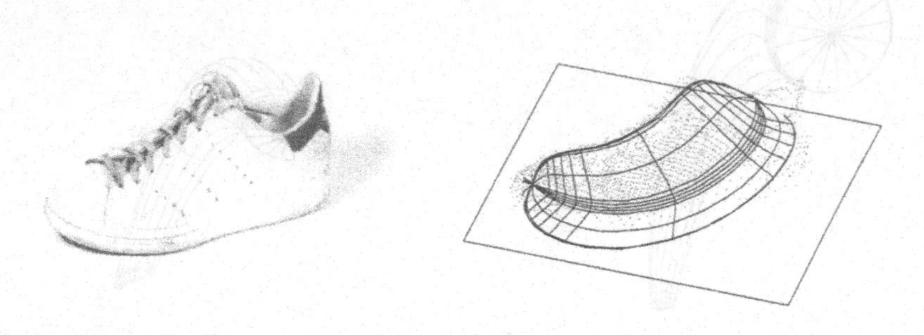

Figure 3.7. Superquadric model of a tennis shoe. The recovered superquadric model which is enhanced with global tapering and bending is sufficient for manipulation tasks but the lack of any details prevents detailed recognition.

details which require further local degrees of freedom. Such enhancement of modeling capabilities offers the possibility of decoupling local (nonrigid) shape components from global shape components and global (rigid) motion. These extensions not only increase the expressive power of the superquadric parametric representation, but also give the potential to make them physics-based or "active."

For a more detailed modeling of shapes, several different ways of introducing local deformations to superquadric models were proposed. Local deformations which require a much larger set of parameters can model a large variety of natural and biological shapes up to a desired accuracy, even human faces (Pentland and Sclaroff, 1991).

Since the seminal work on deformable models (Terzopoulos et al., 1988a), there has been a considerable interest in using local deformations in computer vision. Their shape model was a symmetry seeking tube model whose shape and motion (translation and rotation in 3D) were modeled by the Lagrangian equation of motion incorporating internal (function of the model's physical shape) and external forces (derived from image data). This model was very promising in modeling local or nonrigid motion and shape, but did not encode global (or rigid) shape and motion parameters. All the deformations and motion were accounted for by local deformation of the surface. Clearly, a model that can decouple global parameters of motion (translation and rotation) and global parameters of shape (tapering, bending, twisting) from local or nonrigid deformations is better suited for vision applications. Such a "hybrid" model can incorporate local deformations into any parametric

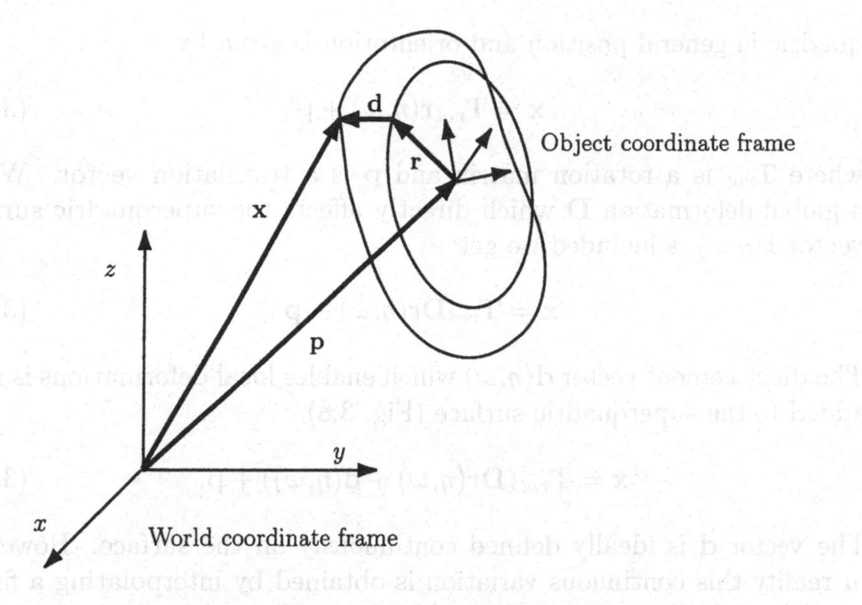

Figure 3.8. Displacement-based hybrid model for adding local deformations (displacement **d** to the surface vector **r**) to superquadric models.

global shape like superquadrics, quadrics, generalized cylinders, algebraic surfaces, etc.

In general, superquadrics can be enhanced with local deformations in four different ways:

- tessellating the surface and adding local degrees of freedom to the discrete mesh resulting in *displacement-based models* which lead to *active models* and *modal representation*,
- extending the superquadric explicit equation by replacing superquadric parameters with parametric functions,
- blending multiple models, and
- using free-form deformations.

Modal representation is related to physics-based models and is derived from a particular way of describing physics-based models (Pentland and Williams, 1989a). Modal representation is discussed in the next chapter together with physics-based reconstruction of models.

3.2.1 DISPLACEMENT-BASED MODELS

Displacement-based or hybrid models augment the superquadric surface vector $\mathbf{r}(\eta, \omega)$ by adding a displacement vector $\mathbf{d}(\eta, \omega)$. A superquadric in general position and orientation is given by

$$\mathbf{x} = \mathbf{T}_{rot}\mathbf{r}(\eta, \omega) + \mathbf{p} , \tag{3.27}$$

where $\mathbf{T}_{rot}$ is a rotation matrix and $\mathbf{p}$ is a translation vector. When a global deformation $\mathbf{D}$ which directly affects the superquadric surface vector $\mathbf{r}(\eta, \omega)$ is included we get

$$\mathbf{x} = \mathbf{T}_{rot}\mathbf{D}\mathbf{r}(\eta, \omega) + \mathbf{p} . \tag{3.28}$$

The displacement vector $\mathbf{d}(\eta, \omega)$ which enables local deformations is now added to the superquadric surface (Fig. 3.8)

$$\mathbf{x} = \mathbf{T}_{rot}(\mathbf{D}\mathbf{r}(\eta, \omega) + \mathbf{d}(\eta, \omega)) + \mathbf{p}. \tag{3.29}$$

The vector $\mathbf{d}$ is ideally defined continuously on the surface. However, in reality this continuous variation is obtained by interpolating a finite number of displacement vectors placed at nodes on the discretized surface (the finite element mesh)

$$\mathbf{d} = \Phi\mathbf{q}_{\mathbf{d}}, \tag{3.30}$$

where Φ contains nodal interpolation functions and $\mathbf{q}_{\mathbf{d}}$ are the local degrees of freedom. A superquadric model can be discretized along the latitude and the longitude to give a tessellated surface. The interpolation is typically achieved by using finite element bilinear or bi-quadric interpolation functions.

Vemuri and Radisavljevic (Vemuri and Radisavljevic, 1993) introduced a multi-resolution dynamic modeling scheme that uses the hybrid model (Eq. 3.29) in an orthonormal wavelet basis (Mallat, 1989). The wavelet basis allows smooth transition between the compact global representation and the distributed local distribution. Their hybrid model has the unique property of being able to smoothly scale up or down the range of possible deformations and the number of parameters required to characterize them. The vector $\mathbf{d}$ in equations (3.29) and (3.30) is augmented as

$$\mathbf{d} = \Phi\mathbf{S}\mathbf{q}_{\mathbf{d}}, \tag{3.31}$$

where S is the discrete wavelet transform. The continuous transformation characteristic generates fractal surfaces of arbitrary degree. Furthermore, the multi-resolution shape models are embedded in a probabilistic framework, allowing model recovery to use a priori information. This model was used for segmenting anatomical structures of the human brain in magnetic resonance images (Vemuri and Radisavljevic, 1993).

3.2.2 PHYSICS-BASED MODELS

The surface discretization and the use of interpolation functions open up the possibility of using the techniques from the field of continuum mechanics and finite element analysis (FEM) which model dynamics of a material's rigid and nonrigid deformation. The terms "physics-based" or "active" models therefore come from this framework. The machinery of FEM allows the continuous material properties like mass and stiffness to be integrated over the entire structure, and formulates the *motion equation* of a general 3D body under influence of external forces (Bathe, 1982)

$$\mathbf{M}\ddot{\mathbf{U}}(t) + \mathbf{C}\dot{\mathbf{U}}(t) + \mathbf{K}\mathbf{U}(t) = \mathbf{R}(t), \tag{3.32}$$

where $\mathbf{M}, \mathbf{C}$ and $\mathbf{K}$ are mass, damping, and stiffness matrices of the body, respectively. The matrices have the form $3n \times 3n$ where n is the number of nodes. $\mathbf{U}$ is a $3n \times 1$ displacement vector, same as $\mathbf{d}$ in equation (3.29), denoting displacements of the body from the unloaded configuration. $\mathbf{R}$ is a $3n \times 1$ vector of the three components of the externally applied forces acting on the nodes. The first term on the left hand side of the equilibrium equation corresponds to the inertial forces, while the second term represents velocity dependent damping forces. The solution to the motion equation (3.32) gives the displacement vectors at equilibrium.

For a linear or nonlinear elastic continuum, the equilibrium solution also satisfies the stationarity condition ($\delta\Pi = 0$) of the total potential energy (Π) of the system. In other words, the equilibrium state (Eq. 3.32) corresponds to the minimum energy state of the system (Bathe, 1982). Consequently, one can formulate the physics-based recovery of a shape by using the variational formulation of the total potential energy, or by directly writing the governing equilibrium equations. Generally speaking, the stationarity condition of the energy results in the *static* version of equation (3.32)

$$\mathbf{K}\mathbf{U}(t) = \mathbf{R}(t), \tag{3.33}$$

which, depending on the problem, is then made dynamic by adding the inertia and the damping terms. Force-based formulations bring in a more physical insight into the problem formulation and allow intuitive understanding of dynamic behavior of the system. For this reason, the physics-based recovery of superquadrics consistently uses a direct formulation, as opposed to a variational one.

All the different types of forces involved in a force-based formulation are time-dependent. However, since a majority of vision applications make the linear elasticity assumption and are concerned only with the final state of the model, a study of the time evolution of the model is not necessary. In such cases, it might be useful to drop the inertia and damping terms and solve equation (3.33) to obtain displacements. Dynamic solutions, with either or both of the inertia and damping terms, are more intuitive for "active" modeling of shapes and for non-rigid motion tracking, and are therefore methods of choice for physics-based superquadric models. The dynamic equation can be seen as establishing a "static" equilibrium at time t, with the effect of inertia and/or damping forces included. Similarly, the static equilibrium equation can then be seen as a special case of "dynamic" equation, with the effect of inertia and damping forces excluded.

In spite of the common theme of using FEM techniques for surface representation and reconstruction by various researchers, there are major differences in types of finite elements, modeling shapes, their recovery, force computation, model-data correspondence, and applications.

3.2.2.1 HYBRID SURFACE MODELS

All the degrees of freedom in the explicit superquadric equation can be directly made "active" (Terzopoulos and Metaxas, 1991). Their hybrid model is composed of both global and local components (Eq. 3.29). They further added global deformations like tapering and bending to the rigid component (Metaxas and Terzopoulos, 1993). Rotation is specified by quarternions. They collected all the degrees of freedom of the model in a vector $\mathbf{q}$, which is made dynamic by introducing mass, damping, and a deformation strain energy of a membrane ($\mathbf{U}$ is replaced by $\mathbf{q}$ in Eq. 3.32). Bilinear triangular finite elements are used to tessellate the superquadric surface. The mass density $\mathbf{M}$ is set to zero to simplify the solution, which is obtained by employing the forward Euler method. Forces are derived from intensity images or from range data. Intensity forces are derived from the image gradient, while long-range model-dependent forces are derived as a scaled distance of the model from data. Again, data-model correspondence is crucial for attaching data forces to model nodes. Minimum distance correspondence is used in the experimental results. A major advantage of their approach is that it allows physics-based recovery of *all* the degrees of freedom of the model. At the same time, their model can use a broad class of finite elements and can be extended to constrained multibody motion for graphics (Metaxas and Terzopoulos, 1992) and vision (Metaxas and Terzopoulos, 1993) applications.

This method was extended to quadratic interpolation functions with thin-plate strain energy and balloon forces to segment left ventricular chamber in dense CT data (McInerney and Terzopoulos, 1993). Qualitative constraints were used to fit globally deformable models to image data under perspective and orthographic projections (Metaxas and Dickinson, 1993).

The hybrid-surface representation was generalized (Metaxas et al., 1997) by doing first a coarse globally deformable superquadric fit on a given set of range points and solving for translation, rotation and global deformation parameters using the physics-based approach (Metaxas and Terzopoulos, 1993). Then, several passes of mesh refinement were done over the coarse superquadric model by locally subdividing triangles, based on the distance between the given datapoints and the model.

The problems of model-to-data correspondence and nonrigid deformation can be solved by a geometric correspondence schema and a dynamic nonrigid deformation model for recovering and tracking 3D data (Gupta and Liang, 1993). The shape model is represented by equation (3.29), where the global degrees of freedom can be recovered by gradient methods or physics-based method (Terzopoulos and Metaxas, 1991). The local deformation vector, $\mathbf{d} = \Phi \mathbf{q_d}$ is however recovered using an iterative scheme that distributes data forces by Gaussian averaging. Thus, the model stiffness, whose role is to keep the surface smooth, is not explicitly modeled but is simulated by force distribution. The model is identical to the forward Euler time-step method for solving the equations of motion, with the mass density $\mathbf{M}$ and stiffness $\mathbf{K}$ set to zero. The model is driven by data-dependent residual forces which vanish (or become negligible small) at equilibrium. This kind of deformable model is suitable only for approximating and simulating observed nonrigid motion. Radial, minimum distance, and viewing direction-based data-model correspondence schema using superquadric models are discussed by the authors. These correspondences are liable to be misaligned in cases where significant concavities exist in the model.

The deformable model (Terzopoulos and Metaxas, 1991) was extended to directly segment the anisotropic volume MRI and ultrafast electron-beam CT images (Gupta et al., 1994). The edge potential was used to generate short-range forces attracting the model to the left-ventricular inner wall (endocardium). The adaptive subdivision of the mesh, which is crucial for obtaining adequate resolution, was performed whenever element edges exceeded a predetermined length. This has the problem of introducing more nodes than locally necessary. The mesh was endowed with orientation, so as to enforce 3D edge-normal constraint to disambiguate close anatomical structures (such as the two edges of the inter-ventricular septum). Mesh orientation was preserved during the subdivision of the mesh. Due to the anisotropic nature of 3D medical data, image forces were computed only in (xy)-plane and interpolated in the inter-slice gap to provide a continuous force field to the model. This work highlights some of the solved and some as yet unsolved problems in 3D segmentation of volume studies using hybrid surface models.

3.2.3 SUPERQUADRICS WITH PARAMETRIC FUNCTIONS

Park, Metaxas and Young (Park et al., 1994) generalized the superquadric equation by replacing the global shape $(\varepsilon_1, \varepsilon_2)$, global size (a_1, a_2, a_3), and the global deformation parameters by "parametric functions" of η. They further added an axis-offset function to the global shape to allow an arbitrary curved axis. These functions increase the degrees of freedom of the model tremendously, since the model is an amalgamation of a "different" superquadric at each η in a continuous manner.

Local deformations were added to the global superquadric surface and the hybrid "parametric" model is recovered using the physics-based approach (Terzopoulos and Metaxas, 1991). The model was applied to the reconstructed displacement field obtained from tagged magnetic resonance data of the left ventricle of the heart (Young and Axel, 1992). In practice, the parametric functions are approximated piecewise at discrete values of η. The authors later extended their model (Park et al., 1995) to describe the wall thickness as a parametric function. This volumetric model is capable of directly deriving forces from spatial modulation of magnetization (SPAMM) correspondence points.

The above hybrid "parametric" superquadrics do not have implicit or closed form equations which makes their recovery and display more difficult. To address this deficiency Zhou proposed to extend superquadrics with exponent functions (Zhou and Kambhamettu, 1999). Extended superquadrics enhance the superquadric shape exponents ε_1 and ε_2 to functions of longitude angle η and latitude angle ω respectively

$$\mathbf{r}(\eta,\omega) = \begin{bmatrix} a_1 \cos^{f_1(\eta)} \eta \cos^{f_2(\omega)} \omega \\ a_2 \cos^{f_1(\eta)} \eta \sin^{f_2(\omega)} \omega \\ a_3 \sin^{f_1(\eta)} \eta \end{bmatrix}, \quad \begin{array}{c} -\pi/2 \le \eta \le \pi/2 \\ -\pi \le \omega < \pi \end{array} . \tag{3.34}$$

Zhou used Bezier curves as exponent functions to make them smooth and controllable. By adjusting the control points of the Bezier curves, one can control the shape of the extended superquadric model. In this way, one can model a broad range of shapes which are not symmetric. Such extended superquadrics also have an implicit equation

$$\left(\left(\frac{x}{a_1}\right)^{\frac{2}{f_2(\arctan \frac{y}{x})}} + \left(\frac{y}{a_2}\right)^{\frac{2}{f_2(\arctan \frac{y}{x})}}\right)^{\frac{f_2(\arctan \frac{y}{x})}{f_1\left(\arctan \frac{z}{\sqrt{x^2+y^2}}\right)}} + \left(\frac{z}{a_3}\right)^{\frac{2}{f_1\left(\arctan \frac{z}{\sqrt{x^2+y^2}}\right)}} = 1. \tag{3.35}$$

Using this equation as the inside-outside function, Zhou defined a recovery method following our approach (Solina and Bajcsy, 1990) and succeeded in recovering extended superquadrics from pre-segmented range images.

A "thick" ventricular model with local deformations instead of parameter functions was also proposed (O'Donnell et al., 1995). The superquadric parameter values can change along the length and thickness of the ventricle.

3.2.4 BLENDED SUPERQUADRIC MODELS

Another method of extending the superquadric vocabulary comes from computer graphics and is called *blending*. Blended superquadric models (DeCarlo and Metaxas, 1998) are created by merging two superquadrics $\mathbf{s}_1$ and $\mathbf{s}_2$ using a blending function $\alpha(\eta)$

$$\mathbf{s}(\eta,\omega) = \mathbf{s}_1(\eta,\omega)\alpha(\eta) + \mathbf{s}_2(\eta,\omega)(1-\alpha(\eta)). \tag{3.36}$$

The blending function $\alpha(\eta)$ is implemented as a nonuniform quadratic B-spline function. Their method can be applied to blending of supertoroids with superquadrics, allowing genus change (0 to 1) during the model recovery. The blending technique allows also non-axial general "gluing" of two superquadric models. The global deformations are applied individually to each of the two models, and the blended model is recovered by the physics-based scheme (Terzopoulos and Metaxas, 1991). The physics-based recovery of blended models is successful also in combination with segmentation of range data (DeCarlo and Metaxas, 1998).

3.2.5 FREE-FORM DEFORMATIONS

Another extension to the superquadrics can be extended with free-form deformations (FFD) (Bardinet et al., 1994). Free-form deformations were first introduced in computer graphics (Sederberg and Parry, 1986). The physical analogy of FFD is to consider a box of flexible plastic in which the superquadric surface is embedded. Each face of the box has points on a regular grid. As these control points are moved, so do all the points in the box, including the superquadric surface. Mathematically, the FFD are defined in terms of a tensor product trivariate Bernstein polynomial. The deformed superquadric surface $\mathbf{X}$ is given by $\mathbf{X} = \mathbf{BP}$, where $\mathbf{B}(s,t,u)$ deforms a point $p(s,t,u)$ by Bernstein polynomial, and $\mathbf{P}$ is a vector containing the control point locations. One of the advantages of FFD is that a parametric surface remains parametric after deformation.

3.3 HYPERQUADRICS

Hyperquadrics are a generalization of superquadrics (Hanson, 1988). By introducing equations with an arbitrary number of linear terms raised to powers, shapes can be generated whose bounding polytopes have an arbitrary number of faces. The basic hyperquadric equation is

$$H(\mathbf{x}) = \sum_{a=1}^{N} \sigma_a |H_a(\mathbf{x})|^{\gamma_a} = C, \tag{3.37}$$

which is a linear combination of $N \leq D$ linear, independent functions $H_a(\mathbf{x})$. Functions $H_a(\mathbf{x})$ are of the form

$$H_a(\mathbf{x}) = \left(\sum_{i=1}^{D} r_{ai} x_i + d_a \right), \tag{3.38}$$

where $\mathbf{x}$ is a D-dimensional vector while r_{ai} and d_a are constants. By taking hyperslices of these high-dimensional algebraic hypersurfaces one can reduce them to the dimension of interest (2D or 3D). Superquadrics are just a special case of hyperquadrics. Hyperquadrics can describe much more complex shapes which do not have to be symmetrical as superquadrics are. Hyperquadrics also have a hierarchical property in the sense that the first terms define the global shape while the following additional terms can introduce more details. Globally deformed hyperquadric shapes typically require up to $N = 5$ terms in equation (3.37). Therefore, deformations which must be explicitly defined for superquadrics can be already incorporated into hyperquadrics.

3.4 RATIOQUADRICS

To enhance the variety of shapes and to address some weaknesses of superquadrics ratioquadrics were proposed as an alternative model (Blanc and Schlick, 1996). They identified the following drawbacks of superquadrics:

- the computation of the signed power function in superquadric equations is rather expensive,

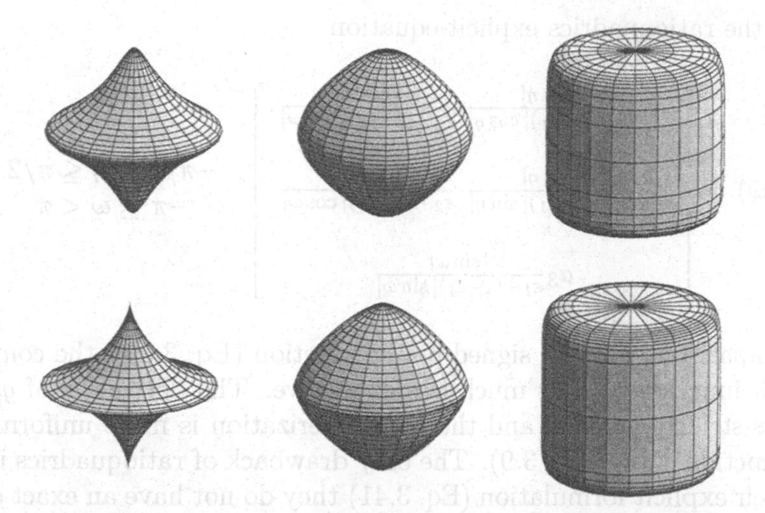

Figure 3.9. Ratioquadrics (Blanc and Schlick, 1996) offer the same power as superquadrics but are faster to compute and behave better in singular points—note the smoother shapes of ratioquadrics (top row) in comparison with superquadrics (bottom row).

- for very small and very large ε_1 and ε_2 the high flatness or steepness of the function when the arguments ($\cos\omega$, $\sin\omega$, $\cos\eta$, $\sin\eta$) of the signed power function are 0, creates numerical imprecision which causes problems, in particular for graphics applications,

- pinchy shapes, which result when ε_1 or $\varepsilon_2 < 1$, have only C^0/G^0 continuity because they have singular points where the first derivative of the superquadric function becomes null.

They proposed solving all of the above drawbacks by replacing the signed power function with a *signed linear rational polynomial* in the range $[-1, 1]$. Replacing the signed power function

$$\forall t \in [-1,1] \quad f_p(t) = \text{sign}(t)|t|^p \tag{3.39}$$

in equation (2.10) with the signed linear rational function

$$\forall t \in [-1,1] \quad g_p(t) = \text{sign}(t)\frac{|t|}{p+(1-p)|t|} \tag{3.40}$$

gives the ratioquadrics explicit equation

$$\mathbf{r}(\eta,\omega) = \begin{bmatrix} a_1 \frac{|\cos\eta|}{\varepsilon_1+(1-\varepsilon_1)|\cos\eta|} \frac{|\cos\omega|}{\varepsilon_2+(1-\varepsilon_2)|\cos\omega|} \\ a_2 \frac{|\sin\eta|}{\varepsilon_1+(1-\varepsilon_1)|\sin\eta|} \frac{|\cos\omega|}{\varepsilon_2+(1-\varepsilon_2)|\cos\omega|} \\ a_3 \frac{|\sin\omega|}{\varepsilon_1+(1-\varepsilon_1)|\sin\omega|} \end{bmatrix} \quad \begin{matrix} -\pi/2 \leq \eta \leq \pi/2 \\ -\pi \leq \omega < \pi \end{matrix} \qquad (3.41)$$

In comparison with the signed power function (Eq. 3.39), the computation of function $g_p(t)$ is much less expensive. The derivative of $g_p(t)$ is always strictly positive and the parameterization is more uniform than the function $f_p(t)$ (Fig. 3.9). The only drawback of ratioquadrics is that for their explicit formulation (Eq. 3.41) they do not have an exact equivalent implicit formulation as superquadrics do. The authors also tried other replacement functions $g_p(t)$ and obtained quadratic ratioquadrics and multiparameter ratioquadrics, where the shape parameter p is no longer a constant but rather a function of η and ω. In this case, similar shapes can be obtained as with extended superquadrics (Zhou and Kambhamettu, 1999).

3.5 SUMMARY

The expressive power of superquadrics can be increased with global and local deformations. With aid of deformations, superquadrics can better and more closely model various natural and man-made shapes. Global deformations which act on the whole model with just a few global deformation parameters can be tightly integrated with superquadric parameterization and recovery. Global deformations therefore, enhance the generic part-level modeling capability of superquadrics. We defined global tapering and bending of superquadrics which in combination require just four additional parameters.

Local deformations, on the other hand, wrap around the superquadric an additional parametric layer, to enable fine-grained local changes of shape. By their nature, the definition of local deformations requires a larger number of parameters and are therefore more difficult to recover from images. They usually have more degrees of freedom than there are data points, so that the recovery process is underconstrained and therefore the recovered models are not unique.

Hyperquadrics have an enriched parameterization, which by itself, enables modeling of deformed shapes which require global deformations when superquadrics are used. Ratioquadrics, on the other hand, are a computational simplification of superquadrics. Ratioquadrics are important for computer graphics applications because they are less expensive to compute and have smoother shapes.

Chapter 4

RECOVERY OF INDIVIDUAL SUPERQUADRICS

One of the criteria for selection of geometric primitives in computer vision is their *accessibility* (Brady, 1983). Accessibility, which can also be defined as computability of the primitive, is essential since the goal of computer vision is to recover structure from images. This requirement not only constrains the choice of the primitives but imposes certain conditions on the model-recovery procedure as well. For example, the primitives should have local support, so that they can cope with occlusions and self-occlusions. Besides, primitives should balance, according to the requirements of the task, the trade-off between data reduction and faithfulness to measured data. All model based approaches are restricted in the sense that they cannot model everything present in the input data. They should, however, model reliably those structures in the image that are essential for a given task. It is also important that the recovery method signals when the models are inadequate to describe the data, so that a different type of model can be invoked.

Superquadrics satisfy the requirement of accessibility. The problem of recovering superquadrics from images is, in general, an over constrained problem. A few superquadric model parameters are determined from hundreds of image features (range points, surface normals or points on occluding contours). By their parameterization, the superquadrics impose a certain symmetry, and in this way place some reasonable constraints on the shape of the occluded portion of a three dimensional object.

In the first part of the chapter, we give an overview of different methods of superquadric recovery. The second part of the chapter is devoted to the least-squares minimization approach to superquadric recovery which is currently one of the most widely used methods. Namely, we give a detailed exposition of the superquadric recovery method that we have developed (Solina, 1987; Solina and Bajcsy, 1990) and that we use also in combination with our method of segmentation (Leonardis et al., 1997). Most recovery methods in this chapter work under the assumption of pre-segmented data, although some methods (Pentland, 1987) address the problem of segmentation and shape recovery in a tightly interwoven fashion. Segmentation with superquadrics is discussed in the next chapter. In the third part of the chapter, we discuss the recovery of superquadrics in the framework of physics-based modeling.

4.1 OVERVIEW OF SUPERQUADRIC RECOVERY METHODS

Each superquadric recovery method makes use of a unique combination of input data, recovery formulation and measure of how well the recovered superquadrics fit the image data. We can classify different superquadric recovery methods from the following three perspectives:

- input data or image type (range, stereo, intensity images),
- type of solution (analytic, minimization of an objective function, interval-bisection method, point distribution model),
- minimization method and goodness-of-fit function (exhaustive search, least-squares minimization, simulated annealing, genetic algorithms).

4.1.1 INPUT DATA

Various types of images have been considered as input to superquadric recovery methods. The most convenient type of images are range images and other images with dense and explicit 3D information such as the images obtained by modern medical imaging techniques. From pairs of intensity images stereo reconstruction is possible. But the drawback of stereo is a sparse and non-uniform distribution of 3D points. When just a single intensity image is available, it is more difficult to establish the relationship between the data and the model.

Superquadrics can be constructed also from finer-grain volumetric models such as oct-trees, which are in turn derived from quadtrees of three orthogonal views of a scene (Ardizzone et al., 1989).

4.1.1.1 RANGE IMAGES

Range data is the input of choice for the majority of superquadric recovery methods (Pentland, 1987; Solina and Bajcsy, 1990; Gross and Boult, 1988; Hager, 1994; Yokoya et al., 1992; Han et al., 1993) since

3D models can be fitted straightforwardly to 3D range points. Namely, values of 3D points can be directly plugged into superquadric equations.

Range data can be acquired with a variety of techniques, such as time of flight techniques which can use laser or sonar, stereo, depth from focus (Krotkov, 1989), and even mechanical probes. With the exception of mechanical probes, the best spatial accuracy can be achieved with structured light techniques (Skočaj, 1999). Because of self-occlusion and sensor geometry, range images captured from a single viewpoint do not offer complete 3D data, but rather only $2\frac{1}{2}$D data (Maver and Bajcsy, 1993). A special type of range data is tactile information which can be obtained from tactile sensors mounted on a robotic gripper. From a set of 3D points, obtained by a dextrous robotic hand which is equipped with tactile sensors, a superquadric model of the grasped object can be obtained (Allen and Michelman, 1990).

4.1.1.2 STEREO

Recovery of superquadrics from stereo images opens specific problems related to highly nonuniform distribution of depth data. In (Chen et al., 1994b) the range points from feature-based stereo are combined with information from occluding contours to solve the problem of outliers which are due to false stereo matching and are frequent along occluding contours. Metaxas (Metaxas and Dickinson, 1993; Chan and Metaxas, 1994) studied the problem of superquadric recovery from stereo and from multiple views in the framework of qualitative shape recovery (Dickinson et al., 1992a) and physics-based recovery technique (Metaxas and Terzopoulos, 1993).

4.1.1.3 SINGLE INTENSITY IMAGES

An interesting issue is the recovery of superquadrics from single 2D intensity images. Intensity images are normally used for derivation of contours, edges, or silhouettes which serve as input for superquadric recovery (Pentland et al., 1991; Vidmar and Solina, 1992). Edge potentials derived from intensity images can serve as a minimization force for superquadric recovery (Terzopoulos and Metaxas, 1991). Intensity images can also be used in a more direct way by employing the shading information which gives the estimates of the local surface normal.

Vidmar and Solina (Vidmar and Solina, 1992) proposed a two step method for the recovery of superquadrics from 2D contours. First, the position of the superquadric is determined, then the shape and orientation are reconstructed using an iterative least-squares method. The fitting function is based on the explicit superquadric equation (2.10) and the condition that the contour points lie on the surface of a superquadric.

Superquadric parameters	*Actual values*	*Initial estimates*	*Recovered values*
a_1	0.2	1	1.00
a_2	0.5	1	0.88
a_3	0.7	1	0.26
ε_1	0.3	0.1	0.70
ε_2	0.3	0.1	1.63

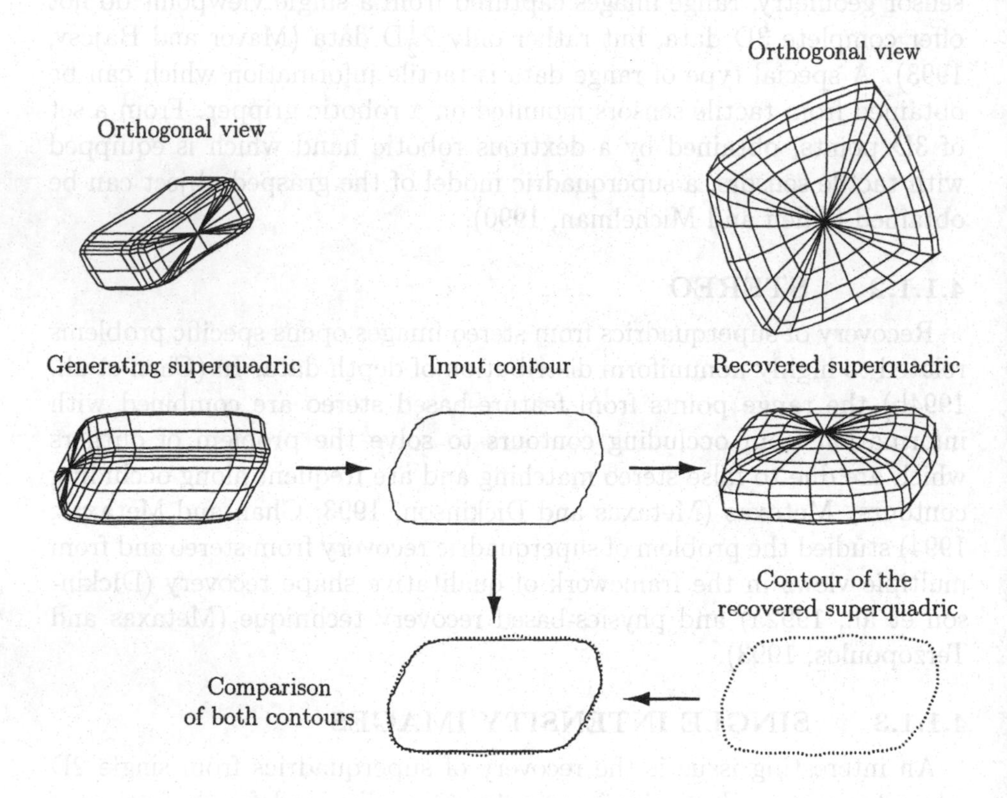

Figure 4.1. Recovery of superquadrics from 2D contours is underconstrained (Vidmar and Solina, 1992). The input contour and the contour of the recovered superquadric closely match. But from the orthogonal views of both superquadrics is quite evident that they are quite different.

For a given contour, often several possible superquadric interpretations are derived. The contour of the recovered model in Fig. 4.1 closely matches the input contour. However, the actual model is quite different, as can be seen from the orthogonal views and also by comparing the actual values of the superquadric parameters with the recovered ones in the Table in Fig. 4.1. To a human observer, some of these interpretations are obviously more natural than others. Perceptually better solutions could be selected by using just a few additional pieces of information (a few range points or shading information).

Dickinson (Dickinson et al., 1992b; Dickinson et al., 1992a) also proposed a method for part model recovery from contours extracted from intensity images which is based on distributed aspect matching. A qualitative shape recovery, which consists of recovering a face topology graph and matching it in a hierarchical way with aspect graphs, can be used first for recognition. For a recovered aspect, there exists a set of possible 3D part primitives, each with a corresponding probability of matching the given image region, which can in turn be used, as proposed in (Metaxas and Dickinson, 1993), for a quantitative shape recovery stage (superquadric models) using physics-based formulation (Terzopoulos and Metaxas, 1991).

Pentland used 2D silhouettes for the segmentation part of his two stage superquadric recovery method (Pentland, 1990). 2D silhouettes of 3D superquadric parts of different shapes, scales and of different orientations were matched to 2D silhouettes in binary images.

Recovery of superquadric models directly from shading information was also attempted (Callari and Maniscalco, 1994). The shading flow field extracted from images was directly matched with the isoluminance field of the approximating shapes. Another method compared the intensity input image directly with synthesized intensity images of superquadric models whose parameters were adjusted by a genetic algorithm (Saito and Tsunashima, 1994; Saito and Kimura, 1996).

4.1.2 TYPES OF SUPERQUADRIC RECOVERY

Different superquadric recovery methods are also distinguished by different types of solutions. The goal of all recovery methods is to find a model that fits the image data and in consequence its parameter values. Most recovery methods define an objective function which can be evaluated for given image data. The goal is to find the parameters of superquadrics that minimize the objective function. Depending on the type of the objective function and the type of input data, different strategies of finding the solution are possible.

4.1.2.1 ANALYTICAL SOLUTION

The first article about superquadrics in computer vision (Pentland, 1986) discussed an analytical solution for recovery of superquadrics. This analytical approach uses the dual relationship between superquadric surface shape (Eq. 2.10) on page 19 and surface normal (Eq. 2.40) on page 28. From these two equations one gets

$$x_n = \frac{\cos^2\eta\cos^2\omega}{x}, \tag{4.1}$$

$$y_n = \frac{\cos^2\eta\sin^2\omega}{y}, \tag{4.2}$$

so

$$\frac{y_n}{x_n} = \frac{x}{y}\tan^2\omega \tag{4.3}$$

or

$$\tan\omega = \sqrt{\frac{yy_n}{xx_n}}. \tag{4.4}$$

We may derive an alternative expression for ω from the superquadric surface vector (Eq. 2.10)

$$x = a_1\cos^{\varepsilon_1}\eta\cos^{\varepsilon_2}\omega, \qquad y = a_2\cos^{\varepsilon_1}\eta\sin^{\varepsilon_2}\omega \tag{4.5}$$

so

$$\frac{x}{y} = \frac{a_1}{a_2}\left(\frac{\cos\omega}{\sin\omega}\right)^{\varepsilon_2} \tag{4.6}$$

or

$$\tan\omega = \left(\frac{ya_1}{xa_2}\right)^{1/\varepsilon_2}. \tag{4.7}$$

Combining the two expression for $\tan\omega$ (Eqs. 4.4 and 4.7) we obtain

$$\frac{y_n}{x_n} = \left(\frac{y}{x}\right)^{2/\varepsilon_2 - 1}\left(\frac{a_1}{a_2}\right)^{2/\varepsilon_2}. \tag{4.8}$$

Letting $\tau = y_n/x_n$, $k = (a_1/a_2)^{2/\varepsilon_2}$ and $\xi = 2/\varepsilon_2 - 1$ we can rewrite the above expression as

$$\tau = k\left(\frac{y}{x}\right)^{\xi}, \tag{4.9}$$

$$\frac{\partial\tau}{\partial y} = \frac{k\xi}{x}\left(\frac{y}{x}\right)^{\xi-1}, \tag{4.10}$$

$$\frac{\partial\tau}{\partial x} = \frac{-k\xi y}{x^2}\left(\frac{y}{x}\right)^{\xi-1} \tag{4.11}$$

This gives two equations relating the unknown shape parameters with quantities measurable in images

$$\frac{\tau}{\partial\tau/\partial y} = \frac{y}{\xi} \tag{4.12}$$

and

$$\frac{\tau}{\partial \tau / \partial x} = \frac{-x}{\xi}. \tag{4.13}$$

Based on equations (4.12) and (4.13) Pentland suggested the construction of a linear regression for solving the center and orientation of the model, as well as the shape parameter ε_2 given, one can estimate the surface-tilt direction τ. From surface-tilt measurements based on shading, texture and/or contour information the model parameters should be computable by means of linear regression. Nevertheless, except for some simple synthetic images, this analytical approach was not successful.

4.1.2.2 MINIMIZATION OF AN OBJECTIVE FUNCTION

The idea behind all minimization approaches is to find a suitable cost, that is a fitting or an objective function which can be evaluated for all considered data points. When the sum of evaluations for all data points is minimized, the parameters (shape, size, orientation, position) of the best fitting superquadric are revealed. A further twist in this problem is due to self-occlusion. Surface data points which are acquired by a single sensor (camera or range scanner) are normally available only on one side of an object and therefore additional constraints must be employed to find a unique superquadric model.

A "brute force" way of finding the minimum of an objective function is to search the entire parameter space. Pentland proposed such a coarse search through the entire superquadric parameter space (Pentland, 1987). A goodness-of-fit functional is evaluated at selected points in the parameter space for many largely overlapping range image regions. The goodness-of-fit function is simply the number of pixels whose range is within $\pm\sigma$ of the model's range, minus λ times the number of pixels that lie off the figure entirely. The model-data correspondence is along z axis. This segmentation stage is later followed by a gradient-descent optimization for individual superquadric models. The main drawback of this method is its excessive computational cost.

From a computational point of view, iterative gradient minimization methods are much more desirable, where the minimum of an appropriate objective function can be found in a small number of iterations. Gradient minimization methods and different objective functions will be discussed more thoroughly in the second part of this chapter.

4.1.2.3 INTERVAL-BISECTION METHOD

Hager studied the problem of sensor-based decision making (Hager, 1994). He was interested in the combination of the estimation (recovery) process with the decision-making process to answer questions like, is an object graspable with a given robotic manipulator. Usually, both processes are separated in the sense that first a shape recovery process is performed and then a decision is made based on the recovered models. Convergence on complex and non-linear problems are difficult to ensure on the one hand, and the amount of computation spent on the fitting stage may be inappropriate for the decision sought on the other hand. Combining both stages should result in minimal work required to reach a decision. Hager demonstrates the approach on two different problems; graspability and categorization, using range data and superquadric models. The approach is based on a interval-bisection method to incorporate sensor-based decision making in the presence of parametric constraints. The constraints describe a model for sensor data (i.e., superquadric) and the criteria for correct decisions about the data (i.e., categorization). An incremental constraint solving technique performs the minimal model recovery required to reach a decision. To keep it computationally feasible, the method does not augment the position and orientation parameters during fitting. The major drawback of the method is slow convergence when categorization is involved. Determination of the shape parameters ε_1 and ε_2 of a superquadric requires several hundred iterations.

4.1.2.4 POINT DISTRIBUTION MODEL

Pilu (Pilu et al., 1996) proposed a method for recovery of deformed superellipses based on the Point Distribution Model (PDM) which can be extended also to superquadrics. The PDM (Cootes et al., 1992) is a statistical finite-element model which is built from a training set of labeled contour landmarks of a large number of shape examples. Although PDMs are normally trained on actual shape instances, Pilu trained them on a large number of randomly generated deformed superellipses. For deformations, he used our definition of linear tapering and circular bending (Solina and Bajcsy, 1990). For PDM training, the points must be labeled and put in correspondence across the whole training set. The mean shape is calculated by averaging each coordinate point and the covariance matrix of the points is computed. The eigenvalue decomposition of the covariance matrix enables the approximation of any shape in the training set by using a weighted sum of the most significant eigenvectors. The weights are called *modes of variations* and they form a new parametric model which can not only represent shapes from the training set, but also new shapes which are in the range of the training set. Pilu used only the first seven modes to represent deformed superellipses. The first four modes are highly correlated to the size of the superellipse (a_1, a_2) and to the bending and tapering parameter. The shape parameter ε

is not strongly correlated with any of the modes. This can be explained by the fact that ε does not affect any major structural change of the superellipse shape. Pilu used such models in the active shape model framework for part-based grouping and segmentation (Pilu and Fisher, 1996a) which is discussed in Chapter 5.

4.1.3 MINIMIZATION METHODS

Minimization methods for superquadric recovery require a definition of a cost, objective or fitting function which can be evaluated for all data points. The superquadric which has the best fit to a given set of surface points has such a size, shape, orientation, and position that the sum of the fitting function values for all points is minimal. All minimization methods used for superquadric recovery must proceed in an iterative fashion since the objective functions are highly non-linear.

Least-squares methods assume that the solution space of the selected objective function is convex enough so that the method eventually comes to an acceptable solution. Simulated annealing and genetic methods on the other hand do not put such severe restrictions on the objective function, but require in general, a much higher number of iterations and a higher computational cost. Different optimization methods are discussed and compared in Section 5.2.3.4 of Chapter 5 where segmentation is also formulated as an optimization problem—searching for a global extremum of a cost function.

Additional constraints for recovery of the perceptually best superquadric model must be used if, due to self-occlusion, range points on only one side of an object are available.

Iterative methods based on non-linear least squares fitting techniques using different distance metrics were proposed soon after the seminal paper by Pentland in 1986 (Bajcsy and Solina, 1987; Boult and Gross, 1987) and are still the most widespread. Least-squares minimization methods for superquadric recovery in combination with different objective functions are discussed thoroughly in Section 4.2 of this Chapter.

4.1.3.1 SIMULATED ANNEALING

Simulated annealing was used to minimize a new error-of-fit measure for recovery of superquadrics from pre-segmented range data (Yokoya et al., 1992). The error-of-fit measure is a linear combination of the distance of range points to the superquadric surface and of the difference in surface normals which was first proposed by (Bajcsy and Solina, 1987). The distance measure itself uses Euclidean distance (Boult and Gross, 1988) which is weighted by the volume factor (Solina and Bajcsy, 1990).

Several hundred iterations were needed to recover models from real range data.

A method for recovery of seven types of parameterized geons which are defined on the basis of superquadrics was proposed by (Wu and Levine, 1994a; Wu and Levine, 1994b). To enable numerical comparison of the fitting residual, they proposed an objective function, made out of two additive terms: a distance measure and a normal vector measure which both reflect the difference in size and shape between object data and parametric models. The distance measure is based on the radial Euclidean distance (Eq. 2.43). For minimization, a special fast simulated annealing technique is used. Still, the reported time required for fitting a single model was between 30 minutes to two hours!

4.1.3.2 GENETIC ALGORITHMS

Intensity input images were directly compared with synthesized intensity images of superquadric models whose parameters were adjusted by a genetic algorithm (Saito and Tsunashima, 1994; Saito and Kimura, 1996). Superquadric parameters are represented with a string which is subjected to genetic operations of crossover and mutation.

4.2 GRADIENT LEAST-SQUARES MINIMIZATION

We were the first who formulated the recovery of superquadric models from pre-segmented range images as a least-squares minimization of a fitting or objective function (Solina and Bajcsy, 1990). The fitting function is based on the inside-outside function F (Eq. 2.16) which is known in the literature also as algebraic distance. To address the non-even distribution of data points due to self-occlusion, we formulated the fitting function so that among all superquadrics that fit the data points, the preference is given to smaller ones. We used an iterative gradient descent method to solve the non-linear minimization problem. A detailed explanation of this method is in Section 4.2.1 of this chapter.

Boult and Gross followed a similar strategy but studied different error-of-fit measures for recovering superquadrics from range data (Boult and Gross, 1988; Gross and Boult, 1988). In particular, they reason that the measure of fit based on the radial Euclidean distance (Eq. 2.43) is a better metric than the algebraic measure based on the inside-outside function F (Eq. 2.16). However, the most intuitive measure—the exact Euclidean distance from a given point to the nearest point on the superquadric surface—can not be expressed analytically and requires iterative numerical methods to be computed (see Section 2.4.2 of Chapter 2). Since the computation of error metrics which are based on the inside-outside function are simpler than the computation of the radial Euclidean distance and since the difference in recovered models is visually negligible—a fact observed also by (Ferrie et al., 1993)—cost functions based on the inside-outside function are used more often. Boult and Gross used the Gauss-Newton method for least-squares minimization. In general, however, fitting with algebraic distance may introduce a *bias*, holes or unbounded surface components (Sullivan et al., 1994).

Whaite and Ferrie addressed the problem of incomplete data due to self-occlusion from a more conservative standpoint (Whaite and Ferrie, 1991). They used the radial Euclidean distance as a measure of error for superquadric recovery, and at the same time did not use any additional constraints besides the ones inherent to superquadrics, that is symmetry and convexity. Using a Gaussian noise assumption, they studied the propagation of errors and formulated a shell of uncertainty that encloses the surface of the fitted model, and in which, there is a given probability that the true surface of the volumetric models lies. The resulting superquadric shells of uncertainty are, in general, larger than the actual objects. In addition, the degree of nonuniqueness of these superquadrics is also much larger since, for example, a planar rectangular patch of range data can be modeled with any superquadric with a flat face large enough to cover the data. This knowledge on where the model's surface has a higher uncertainty can be used for active exploration and planning the next view to enable model validation or to build composite scene models from several views (Whaite and Ferrie, 1997).

The objective function used for minimization was multiplied with a weighting function for robust estimation by (Horikoshi and Suzuki, 1993). The weighting function consists of three forms. Which form is used for evaluation of a particular range point depends on where the point lies relative to the surface of the superquadric model: close to the median value of the inside-outside function, or farther away in either direction. Consequently, the superquadric model is less sensitive to outliers, and the resulting global model hugs the object better than what is obtained by our least squares formulation. The compact volume obtained by their method was crucial for the success of their segmentation algorithm.

Evaluation of different cost functions for gradient least-squares minimization was done by (van Dop and Regtien, 1998). In particular, they compared the performance of our algebraic cost function, which is based on the inside-outside function, and the radial Euclidean distance (Whaite and Ferrie, 1991). They compared superquadrics recovered from several pre-segmented, noisy range images. Initially, they got better results with our cost function. By adding to the radial Euclidean distance method, first, a background constraint and, second, a robust technique for elimination of outliers, they finally got better results, but only for cylindric objects. The background constraint assumes that the superquadric models rest on a planar support plane (Gupta et al., 1989b). Since this background constraint cannot be expressed analytically as a constraint on superquadric parameters, points which are on the object's surface are projected onto the supporting plane. These projected points are assumed as supporting points, which are on the surface of the occluded side of the superquadric. These points are at the moment *inside* of the current superquadric model and are included as surface range points for the next iteration in the minimization process. The robust technique for elimination of outliers is based on the least trimmed squares.

4.2.1 SUPERQUADRIC RECOVERY USING THE INSIDE-OUTSIDE FUNCTION

This section gives a thorough review of the superquadric recovery method which we defined in 1987 (Solina, 1987; Bajcsy and Solina, 1987; Solina and Bajcsy, 1987; Solina and Bajcsy, 1990).

Suppose we have a set of n 3-D surface points expressed in some world coordinate system (x_i, y_i, z_i), $i = 1, \ldots, n$ which we want to model with a superquadric. A superquadric in general position is defined by the following equation (see also Eq. 2.35)

$$F(x, y, z; \lambda_1, \ldots \lambda_{11}) = 1 . \tag{4.14}$$

We want to find such values for the 11 parameters $\boldsymbol{\Lambda}$ ($\lambda_j, j = 1, \ldots, 11$) that most of the n 3-D points (x_i, y_i, z_i) will lay on, or close to the superquadric's surface.

There will probably not exist a set of parameters $\boldsymbol{\Lambda}$ that perfectly fits the data. Finding the model $\boldsymbol{\Lambda}$ for which the algebraic distance from points to the model is minimal, is defined as a least-squares minimization problem. Since, for any point (x, y, z) on the surface of a superquadric holds

$$F(x, y, z; \lambda_1, \ldots, \lambda_{11}) = 1, \tag{4.15}$$

we have to minimize the following expression

$$\min_{\boldsymbol{\Lambda}} \sum_{i=1}^{n} (F(x_i, y_i, z_i; \lambda_1, \ldots, \lambda_{11}) - 1)^2 . \tag{4.16}$$

4.2.1.1 THE PROBLEM OF SELF-OCCLUSION

Due to self-occlusion, not all sides of an object are visible at the same time. Range data in particular, when captured from a single view, is incomplete not only due to self-occlusion, but also as a result of sensor set-up (Maver and Bajcsy, 1993). Range points then cover typically just a part of the imaged objects. Even more, from a particular viewing direction, just one side of a 3D object can be seen. Obviously, from such "degenerate" views the shape of an object can not be recovered (Koenderink and van Doorn, 1979; Kender and Freudenstein, 1987). But even when a "general" viewpoint is assumed, certain objects such as parallelepipeds or cylinders (objects with surfaces where at least one principal curvature equals zero) do not provide sufficient constraints for a unique shape recovery using merely the inside-outside function (Eq. 4.16). For example, a range image of a box which is taken from a single viewpoint, has range points only on three sides of the box. Several superquadric models satisfy equation (4.15). Among all possible superquadrics which fit the range data points in the least squares sense, we want to find the smallest superquadric (Fig. 4.2).

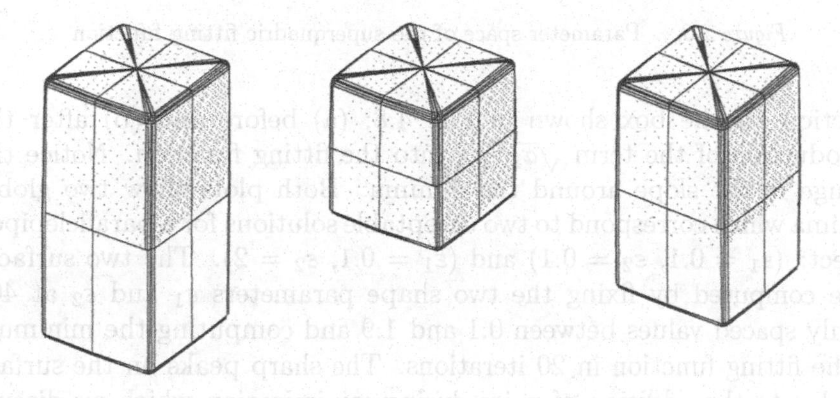

Figure 4.2. **Among all superquadric which are in the least-square sense aligned with the range points we want to find the one with the smallest volume.**

To enforce the recovery of the smallest superquadric that best fits the 3D points in the least squares sense, we introduced the additional constraint by multiplying the minimization terms with $\sqrt{a_1 a_2 a_3}$ which is the same as $\sqrt{\lambda_1 \lambda_2 \lambda_3}$

$$\min_{\Lambda} \sum_{i=1}^{n} \left(\sqrt{\lambda_1 \lambda_2 \lambda_3} \left(F(x_i, y_i, z_i; \lambda_1, \ldots, \lambda_{11}) - 1 \right) \right)^2 . \qquad (4.17)$$

Fig. 4.3 shows how the term $\sqrt{a_1 a_2 a_3}$ affects the shape of the solution space. The two surfaces show part of the parameter space of the fitting

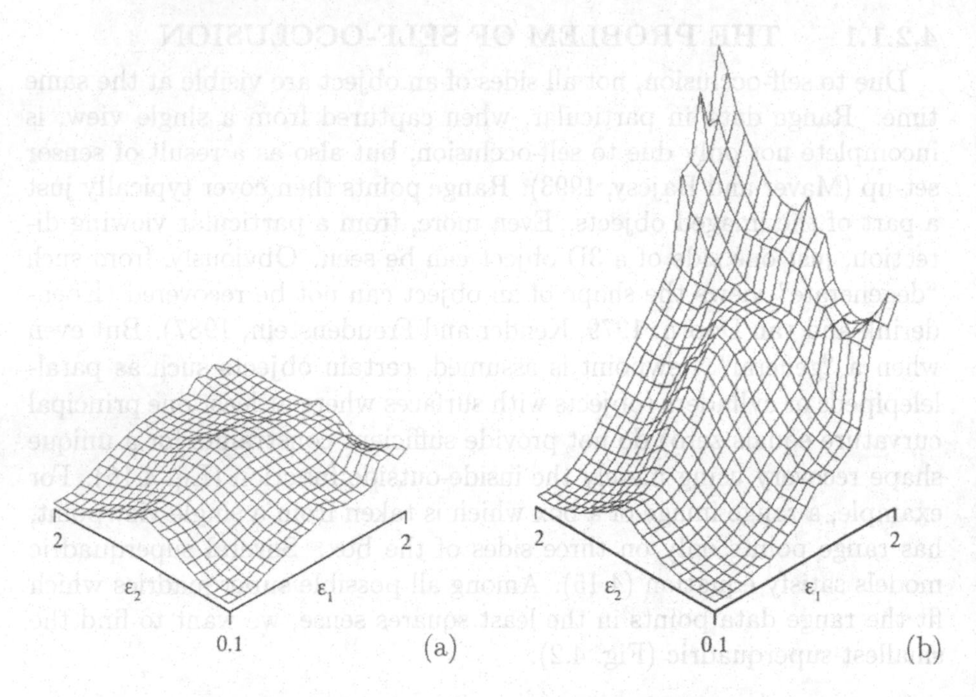

Figure 4.3. Parameter space of the superquadric fitting function

function for the box shown in Fig. 4.8; (a) before and (b) after the introduction of the term $\sqrt{a_1 a_2 a_3}$ into the fitting function. Notice the change in the slope around the minima. Both plots show two global minima which correspond to two acceptable solutions for a parallelepiped object: ($\varepsilon_1 = 0.1$, $\varepsilon_2 = 0.1$) and ($\varepsilon_1 = 0.1$, $\varepsilon_2 = 2$). The two surfaces were computed by fixing the two shape parameters ε_1 and ε_2 at 400 evenly spaced values between 0.1 and 1.9 and computing the minimum of the fitting function in 20 iterations. The sharp peaks on the surface are due to the addition of noise during minimization which we discuss later.

In a similar fashion, the normal measure used by (Wu and Levine, 1994a; Wu and Levine, 1994b) is multiplied with the factor $(a_1 + a_2 + a_3)/3$, which forces the selection of the model of the smaller size when models of different sizes fit the object data points equally well.

Recently, a similarity between the multiplication of our cost function with the superquadric volume and a robust method for fitting ellipses to n scattered data points was observed (Fitzgibbon et al., 1996). The robust direct least-squares method is based on the minimization of a cost function, which is based on the algebraic distance

$$F = ax^2 + bxy + cy^2 + dx + ey + f = 0, \tag{4.18}$$

and incorporates the elliptical constraint $4ac - b^2 = 1$. The minimization problem is directly solvable by an eigensystem. Minimizing

$$\frac{1}{4ac - b^2} \sum_{i=1}^{n} F^2(x, y) \tag{4.19}$$

gives a unique solution and is analogous to multiplying the cost function F with the area of the ellipse since the discriminant $b^2 - 4ac$ is inversely proportional to the product of the radii.

4.2.1.2 THE PROBLEM OF PARAMETER SPACE

An intervention that made the fitting function more suited for rapid convergence during minimization was to raise the inside-outside function F (Eq. 2.16) to the power of ε_1 which made the error metric independent from the shape of the superquadric that is regulated by ε_1 (Fig. 4.4),

$$F^{\varepsilon_1}(x, y, z) = \left(\left(\left(\frac{x}{a_1}\right)^{\frac{2}{\varepsilon_2}} + \left(\frac{y}{a_2}\right)^{\frac{2}{\varepsilon_2}} \right)^{\frac{\varepsilon_2}{\varepsilon_1}} + \left(\frac{z}{a_3}\right)^{\frac{2}{\varepsilon_1}} \right)^{\varepsilon_1} . \tag{4.20}$$

This intervention obviously does not affect the shape of the superquadric itself (when $F = 1$). Even after the introduction of the exponent ε_1, the error metric is still not completely uniform in all directions and depends on the ratio of the size parameters. In the direction of the longer dimension of the superquadrics, the error measure grows slower (Fig. 4.4). Despite this deficiency, our fitting function is efficient and robust in recovering compact superquadric models (Solina and Bajcsy, 1990)[1].

The new fitting function is therefore

$$F_S(x, y, z) = \sqrt{a_1 a_2 a_3}(F^{\varepsilon_1}(x, y, z) - 1), \tag{4.21}$$

where a_1, a_2, a_3 (or $\lambda_1, \lambda_2, \lambda_3$) are the superquadric size parameters. The relationship of the function F_S to the radial Euclidean distance was first pointed out by (Whaite and Ferrie, 1991) and can be derived from Eq. (2.43) in Chapter 2 by multiplying the following two equations

$$F^{\frac{\varepsilon_1}{2}}(x, y, z) - 1 = \frac{d}{|\mathbf{r_S}|}, \tag{4.22}$$

$$F^{\frac{\varepsilon_1}{2}}(x, y, z) + 1 = \frac{d}{|\mathbf{r_S}|} + 2, \tag{4.23}$$

[1] In the same way was adjusted the error-of-fit function for recovery of extended superquadrics (Zhou and Kambhamettu, 1999).

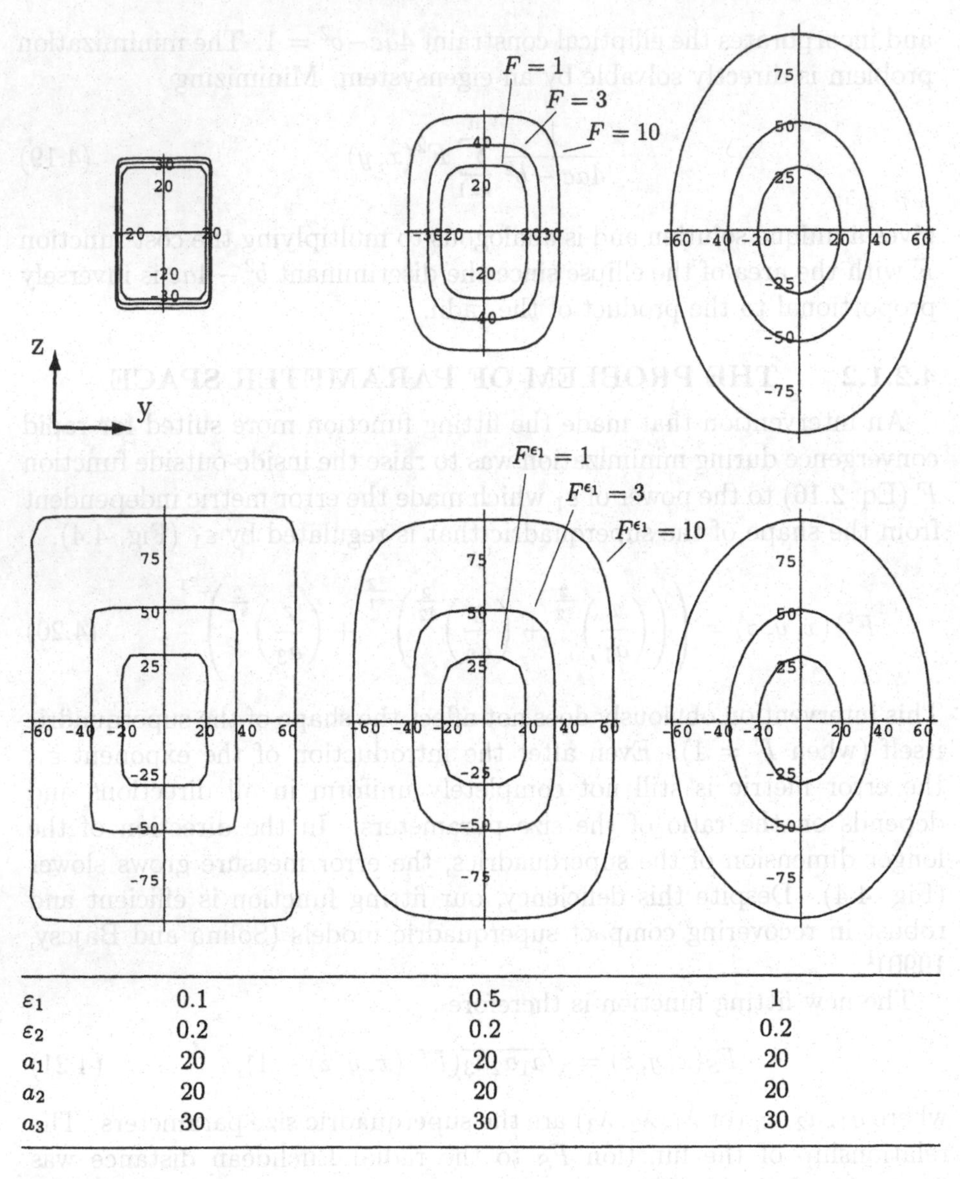

ε_1	0.1	0.5	1
ε_2	0.2	0.2	0.2
a_1	20	20	20
a_2	20	20	20
a_3	30	30	30

Figure 4.4. Cross sections of superquadrics and their inside-outside functions at values $F = 1$, $F = 3$, and $F = 10$ in (yz)-plane. The top row shows the standard inside-outside functions F (Eq. 2.16) and the bottom row after raising F to the power of ε_1 (Eq. 4.20).

resulting in

$$F^{\varepsilon_1}(x, y, z) - 1 = \frac{d}{|\mathbf{r_S}|}\left(\frac{d}{|\mathbf{r_S}|} + 2\right), \tag{4.24}$$

where $\mathbf{r_S}$ is the vector from the superquadric center to the point on its surface from which the radial Euclidean distance d is measured. The metric in form $(F^{\varepsilon_1}(x, y, z) - 1)$ is thus biased towards larger superquadrics that will produce larger values for $|\mathbf{r_S}|$. This effect is partially compensated by the multiplication with the term $\sqrt{a_1 a_2 a_3}$ which is proportional to the volume of the superquadric.

Now, we have to minimize

$$\min_{\Lambda} \sum_{i=1}^{n} \left(\sqrt{\lambda_1 \lambda_2 \lambda_3} (F^{\varepsilon_1}(x_i, y_i, z_i; \lambda_1, \ldots, \lambda_{11}) - 1) \right)^2. \quad (4.25)$$

Since this expression is a nonlinear function of 11 parameters in $\mathbf{\Lambda}$ ($\lambda_j, j = 1, \ldots, 11$), minimization must proceed iteratively. Given a trial set of values of model parameters $\mathbf{\Lambda}_k$, at k-th iteration, we evaluate F_S (Eq. 4.21) for all n points and employ a procedure to improve the trial solution. The procedure is then repeated with a set of new trial values $\mathbf{\Lambda}_{k+1}$ until the fitting function (Eq. 4.25) stops decreasing, or the changes are statistically insignificant. We use the Levenberg-Marquardt method for nonlinear least squares minimization (Press et al., 1986; Scales, 1985), shown in Table 4.1, since the first derivatives of the fitting function $\partial F_S / \partial \lambda_i$ for $i = 1, \ldots, 11$ can be computed analytically.

4.2.1.3 THE PROBLEM OF LOCAL MINIMA

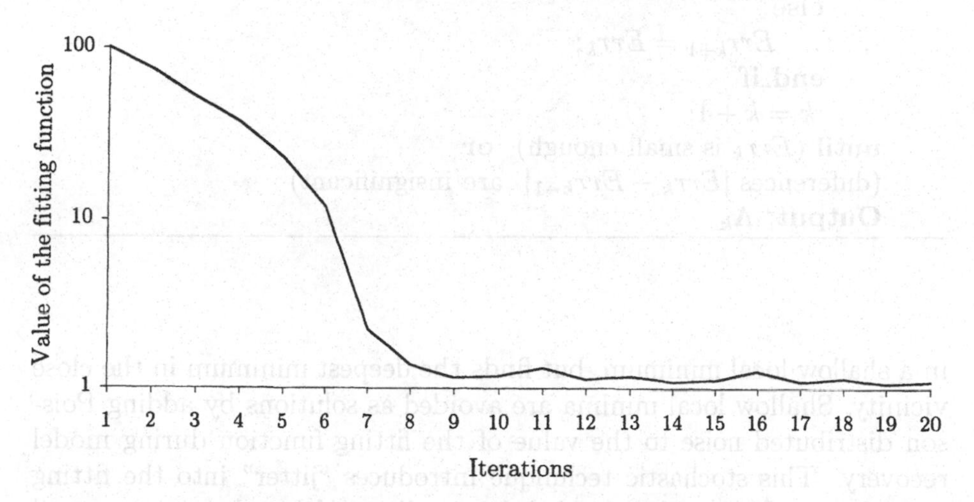

Figure 4.5. The value of the fitting function sharply decreases during shape recovery of the banana shown in Figure 4.10. The jaggedness of the function is due to the addition of Poisson distributed noise during model recovery.

Minimization methods can, in general, guarantee convergence only to a local minimum. To which minimum converges the minimization procedure depends on the starting position in the parameter space ($\mathbf{\Lambda}_0$). We have to assure that the minimization procedure does not get stuck

Table 4.1. Algorithm for iterative superquadric recovery using the Levenberg-Marquardt minimization method for non-linear least squares minimization

Input data:
    set of n 3D range points $\mathbf{x}_{w_1}, \ldots, \mathbf{x}_{w_n}$;
    initial model parameters $\mathbf{\Lambda}_0$;
    fitting function R;
$Err_0 = 0$;
for $j = 1$ **to** n **do**
    $Err_0 = Err_0 + R^2(\mathbf{x}_j, \mathbf{\Lambda}_0)$;
end_for
$k = 0$;
repeat
    Compute $\mathbf{\Lambda}_{k+1}$ using Levenberg-Marquardt method;
    $Err_{k+1} = 0$;
    for $j = 1$ **to** n **do**
        $Err_{k+1} = Err_{k+1} + (\sqrt{\lambda_1\lambda_2\lambda_3}(F^{\varepsilon_1}(x_j, \mathbf{\Lambda}_{k+1}) - 1))^2$;
    end_for
    if $Err_{k+1} < Err_k$ + Poisson noise
        Accept $\mathbf{\Lambda}_{\mathbf{k+1}}$;
    else
        $Err_{k+1} = Err_k$;
    end_if
    $k = k + 1$;
until (Err_k is small enough) **or**
(differences $|Err_k - Err_{k-1}|$ are insignificant)
Output: $\mathbf{\Lambda}_k$

in a shallow local minimum, but finds the deepest minimum in the close vicinity. Shallow local minima are avoided as solutions by adding Poisson distributed noise to the value of the fitting function during model recovery. This stochastic technique introduces "jitter" into the fitting procedure and resembles simulated annealing. Although not as general as simulated annealing, this technique is much faster. Fig. 4.5 shows the changing values of the fitting function during model recovery of the banana in Fig. 4.10.

4.2.1.4 PARAMETER CONSTRAINTS

To limit the possible shapes of superquadrics and to avoid singularities during the minimization, we had to introduce inequality constraints on function parameters $\mathbf{\Lambda}$. When $\varepsilon_1, \varepsilon_2 > 2$, superquadrics have concavities and we decided not to use such models in our shape vocabulary.

Constraints that we use are simple bounds on the parameter values in the form of intervals. A constraint becomes active when a parameter reaches the lower or the upper bound of the allowable interval. We observed that the constraints mostly became active only for those parameters $\mathbf{\Lambda}$ that were not accepted because the step in the direction of the derivative was too large. We conjectured that constraints are necessary not so much to assure convergence to a local minimum, but to achieve numerical stability. For example, when any of the parameters $a_1, a_2, a_3, \varepsilon_1, \varepsilon_2$ equals 0, the superquadric inside-outside equation (2.16) becomes singular.

Parameter constraints are implemented by a simple projection method (Scales, 1985). We take the search vector or the trial parameters $\mathbf{\Lambda}$ generated by the unconstrained minimization technique and project the search vector so that it lies in the intersection of the set of constraint intervals. We use the following constraints: $\{a_1, a_2, a_3\} > 0$ and $0.1 < \{\varepsilon_1, \varepsilon_2\} < 2$. When $\{\varepsilon_1, \varepsilon_2\} < 0.1$, the inside-outside function (2.16) might become numerically unstable, although the superquadric shape stays perceptually almost the same.

Some examples of recovery of prototypical shapes with superquadric models are shown in Figs. 4.6–4.8.

4.2.2 COMPUTATION OF INITIAL ESTIMATES OF MODEL PARAMETERS

An initial estimate of the set of model parameters $\mathbf{\Lambda}_0$ determines to which local minimum the minimization procedure will converge. When testing the iterative model recovery method described in the previous section, we found that only very rough estimates of object's true position, orientation, and size suffice to assure convergence to a local minimum that corresponds to the actual shape. This is important since these parameters can be estimated only from the range points on the visible side of the object and hence, the estimates cannot be very accurate to begin with. Initial values for both shape parameters, ε_1 and ε_2 can always be 1, which means that the initial model $\mathbf{\Lambda}_0$ is always an ellipsoid. This insensitivity towards correct estimation of the two shape parameters was achieved when we added the outermost exponent ε_1 to the inside-outside function F (Eq. 4.20).

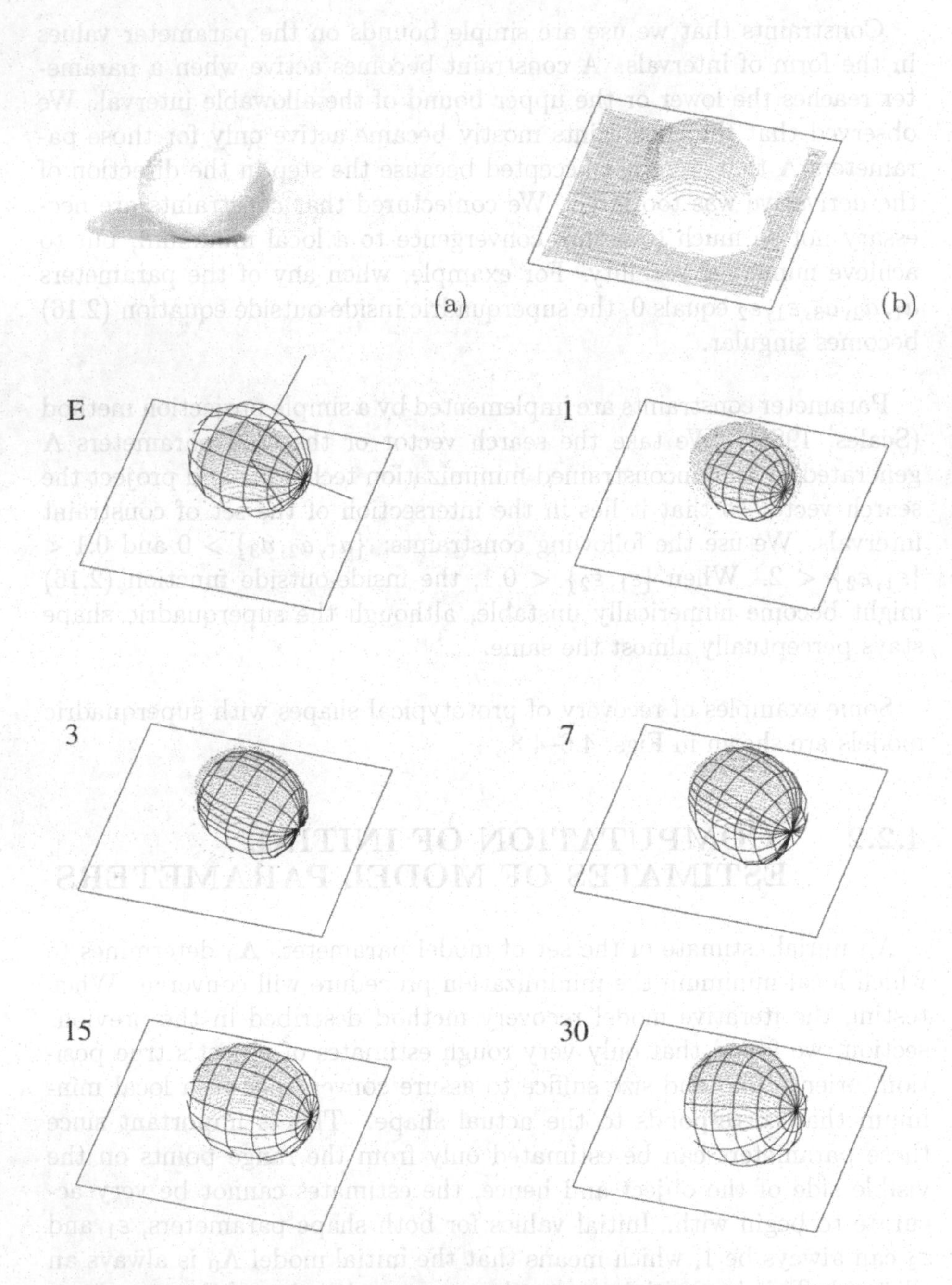

Figure 4.6. **Recovery of a round object: (a) is the intensity image and (b) the corresponding range image. Below is the recovery sequence showing the initial model estimate (*E*) and models after 1st, 3rd, 7th, 15th and 30th iteration.**

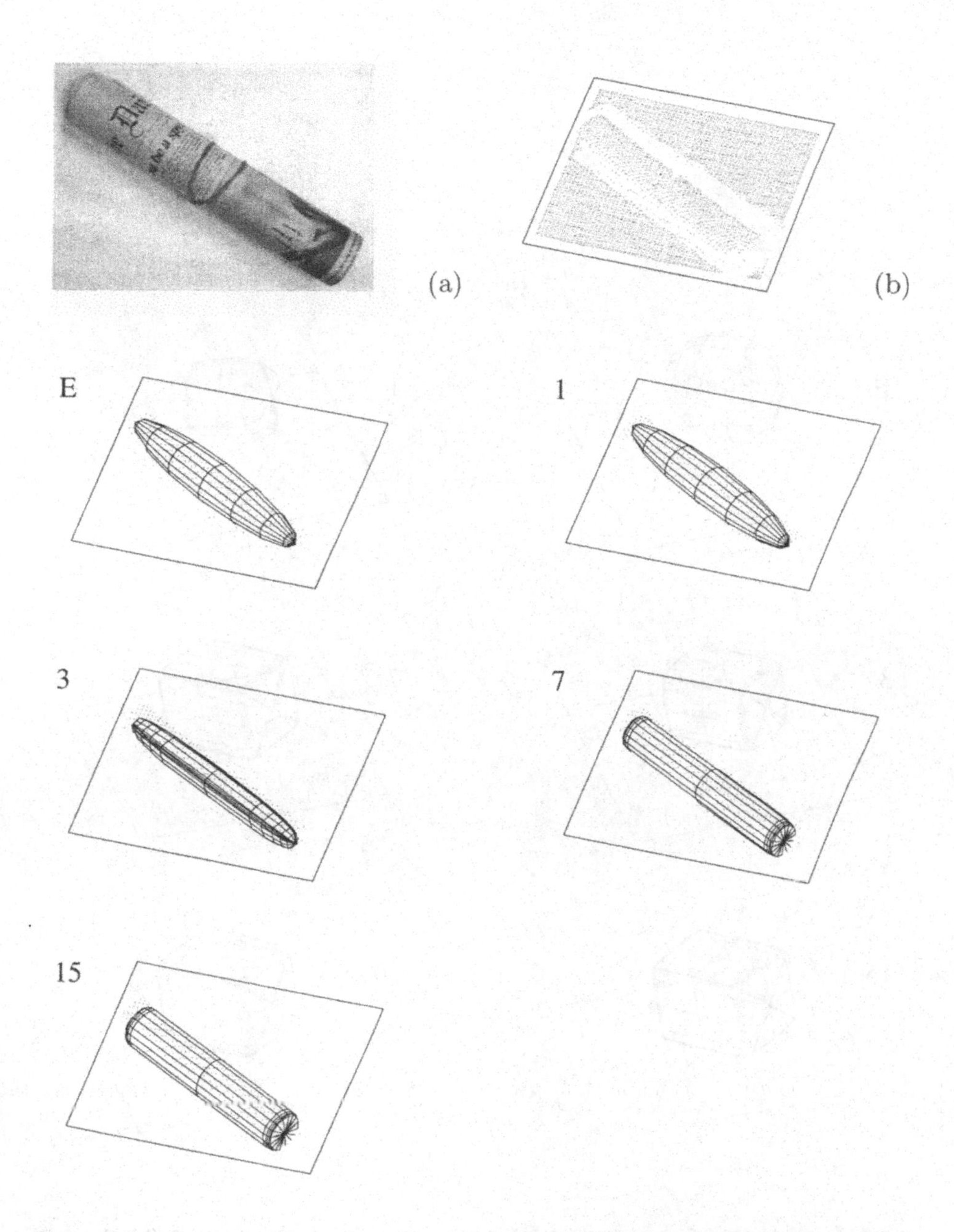

Figure 4.7. Shape recovery of a cylindric object: (a) is the intensity image and (b) the corresponding range image. Below is the recovery sequence showing the initial model estimate (E) and models after 1st, 3rd, 7th and 15th iteration.

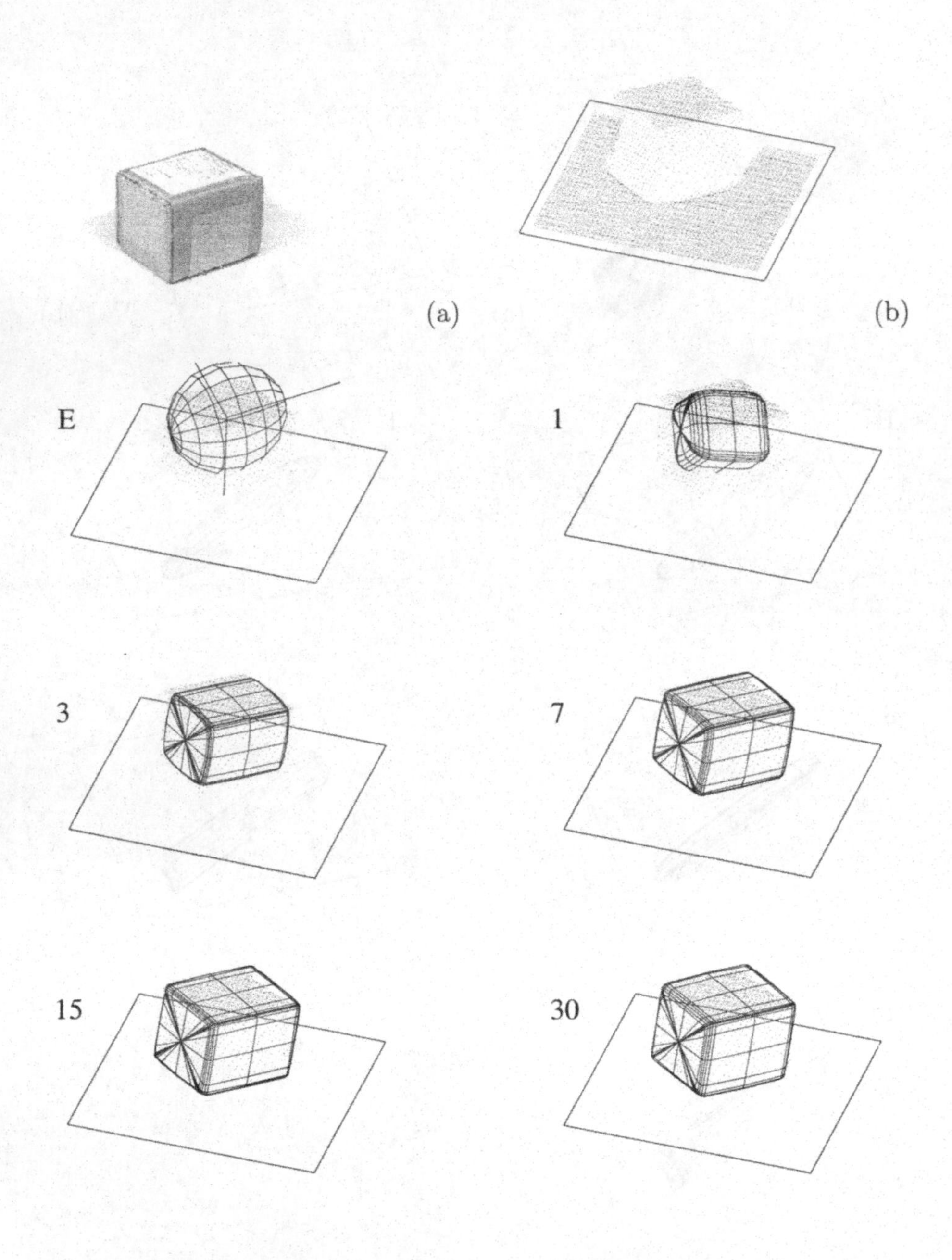

Figure 4.8. Shape recovery of a box: (a) is the intensity image and (b) the corresponding range image. Below is the recovery sequence showing the initial model estimate (E) and models after 1st, 3rd, 7th, 15th and 30th iteration.

The position p_x, p_y, p_y (or $\lambda_9, \lambda_{10}, \lambda_{11}$) of the initial ellipsoid $\mathbf{\Lambda}_0$ is set to the center of gravity of all n range points (x, y, z)

$$p_{x0} = \bar{x} = \frac{1}{n}\sum_{i=1}^{n} x_{w_i} \tag{4.26}$$

$$p_{y0} = \bar{y} = \frac{1}{n}\sum_{i=1}^{n} y_{w_i} \tag{4.27}$$

$$p_{z0} = \bar{z} = \frac{1}{n}\sum_{i=1}^{n} z_{w_i}. \tag{4.28}$$

To compute the orientation of the object-centered coordinate system, we compute first the matrix of central moments

$$\mathbf{M}_C = \frac{1}{n}\sum_{i=1}^{n}\begin{bmatrix} (y_i-\bar{y})^2+(z_i-\bar{z})^2 & -(y_i-\bar{y})(x_i-\bar{x}) & -(z_i-\bar{z})(x_i-\bar{x}) \\ -(x_i-\bar{x})(y_i-\bar{y}) & (x_i-\bar{x})^2+(z_i-\bar{z})^2 & -(z_i-\bar{z})(y_i-\bar{y}) \\ -(x_i-\bar{x})(z_i-\bar{z}) & -(y_i-\bar{y})(z_i-\bar{z}) & (x_i-\bar{x})^2+(y_i-\bar{y})^2 \end{bmatrix}. \tag{4.29}$$

Central moments are moments with respect to the center of gravity $(\bar{x}, \bar{y}, \bar{z})$. We want to find a rotation matrix $\mathbf{T}_R$ which makes the matrix of moments $\mathbf{M}_C$ diagonal (Rosenfeld and Kak, 1982; Horn, 1986). The new diagonal matrix of moments $\mathbf{D}$ is then

$$\mathbf{D} = \mathbf{T}_R^{-1}\mathbf{M}_C\mathbf{T}_R, \tag{4.30}$$

where $\mathbf{T}_R$ is the rotational part of transform $\mathbf{T}$ (2.30) in Chapter 2. On the other hand, matrix $\mathbf{M}_C$ can be diagonalized with a diagonalization matrix $\mathbf{Q}$, whose columns are eigenvectors of matrix $\mathbf{M}_C$ (Strang, 1988)

$$\mathbf{D} = \mathbf{Q}^{-1}\mathbf{M}_C\mathbf{Q}. \tag{4.31}$$

Comparing equations (4.30) and (4.31) gives

$$\mathbf{T}_R = \mathbf{Q}. \tag{4.32}$$

Rotation matrix $\mathbf{T}_R$ that orients the object centered coordinate system along the axis of minimum and maximum inertia, can thus be assembled out of eigenvectors of matrix $\mathbf{M}_C$.

Eigenvector $\mathbf{e}_1$ with the smallest eigenvalue κ_1 corresponds to the minimum-inertia line and the eigenvector $\mathbf{e}_3$ with the largest eigenvalue to the maximum inertia line. The minimum-inertia line is also known as the principal axis (Rosenfeld and Kak, 1982). We use the Jacobian method for computing the matrix of eigenvectors which consists of a sequence of orthogonal similarity transformations designed to annihilate one of the off-diagonal matrix elements. The Jacobi method is fast for matrices of order less than ten and absolutely foolproof for all real symmetric matrices (Press et al., 1986) which is the case with matrix $\mathbf{M}_C$.

We decided to orient the object centered coordinate system so that the new axis z lies along the longest side for elongated objects (axis of least inertia) and along the shortest for flat objects (axis of largest inertia), based on the assumption that bending and tapering deformations normally affect objects along their longest side. For round flat objects, on the other hand, we want the z coordinate axis to coincide with the axis of the rotational symmetry. Given the three eigenvectors $\mathbf{e}_1, \mathbf{e}_2, \mathbf{e}_3$, we have to assign to them coordinate axes labels. As stated above, we want to control only the orientation of the z axis. For ordered eigenvalues $\kappa_1 < \kappa_2 < \kappa_3$, of the three corresponding eigenvectors $\mathbf{e}_1, \mathbf{e}_2, \mathbf{e}_3$, the z axis is assigned according to the following rule

$$\begin{array}{l} \textbf{if } |\kappa_1 - \kappa_2| < |\kappa_2 - \kappa_3| \textbf{ then } \mathbf{z} = \mathbf{e}_3 \\ \textbf{else } \mathbf{z} = \mathbf{e}_1. \end{array}$$

This condition puts the z axis along the longest side of elongated objects and perpendicular for flat, rotationally symmetric objects. From the elements of the rotation matrix $\mathbf{T}_R$ which makes up the rotational part of the homogeneous transformation $\mathbf{H}$ (2.30), we compute the equivalent Euler angles ϕ_0, θ_0, ψ_0 (or $\lambda_6, \lambda_7, \lambda_8$) (Paul, 1981).

For evaluating the inside-outside function (Eq. 2.33) we could use the elements of the rotational matrix $\mathbf{T}_R$ directly, but the partial derivatives required for minimization of the fitting function are all expressed in terms of Euler angles.

The size a_1, a_2, a_3 (or $\lambda_1, \lambda_2, \lambda_3$) of the initial ellipsoid $\mathbf{\Lambda}_0$ are computed from the bounding box of the range points whose sides are aligned with the new object-centered coordinate system.

The initial estimates computed in the described fashion are sometimes very close to the actual parameter values. But even when the initial estimates for rotation are quite poor—this is the case with objects which are not elongated—the correct model is recovered. This would suggest that the estimation of orientation is not critical at all. The initial ellipsoid could have a default orientation, for example, the same as the world coordinate system. This would be acceptable for blob-like objects, but especially for elongated objects we want to align the z coordinate of the object centered coordinate system with the longest axis of the superquadric so that it can be subjected to bending, if necessary.

Recovery of a single superquadric model requires, on the average, only about 20 iterations. The required time for each iteration depends on the number of range points. Since the computations involving individual range points are independent, the method lends itself naturally to a parallel implementation.

Since perfect segmentation is assumed, the method by itself does not cope well with outliers. In the context of the recover-and-select segmentation method which we describe in Chapter 5, the outlier problem is greatly reduced since the inclusion of range points into an individual superquadric model is dynamically adjusted by the segmentation process.

4.2.3 RECOVERY OF DEFORMED SUPERQUADRICS

In this section we describe the recovery of superquadrics enhanced with global deformations which we defined in Chapter 3. Recovery of deformed superquadrics is handled in the same way as the recovery of non-deformed superquadrics. Instead of only 11 parameters, the fitting function has 15 parameters when tapering and bending are used. The initial values for global deformation parameters correspond to non-deformed superquadrics. For tapering (Eq. 3.12), for example, the initial values of parameters $K_{x0} = K_{y0} = 0$. During model recovery, both tapering parameters are adjusted simultaneously with the other 13 parameters (Fig. 4.9).

For bending (Eq. 3.23), the initial values of the bending parameters in the initial set of model parameters $\mathbf{\Lambda}_0$ also correspond to an unbent superquadric. Since, for $k = 0$, equation (3.15) for the bending transformation becomes singular, a very small number is used for k_0 instead. The orientation of the bending plane around axis z is set to $\alpha_0 = 0$, although any real value could be used. During model recovery, both bending parameters are adjusted simultaneously with the other 13 parameters (Fig. 4.10).

Superquadric models can be tapered and bent simultaneously which requires that all 15 parameters are adjusted simultaneously (Eq. 3.26). A recovery sequence of a tapered and bent model is shown in Fig. 4.11.

4.2.4 MULTI-RESOLUTION RECOVERY OF SUPERQUADRICS

The most time consuming part in our superquadric recovery method is the evaluation of the fitting function and all of its partial derivatives for every input range point which must be repeated in each iteration of the minimization procedure. A substantial speedup can therefore be achieved by subsampling the initial set of range points.

During minimization, the value of the fitting function drops very fast until it reaches a plateau. Further iterations gain no substantial im-

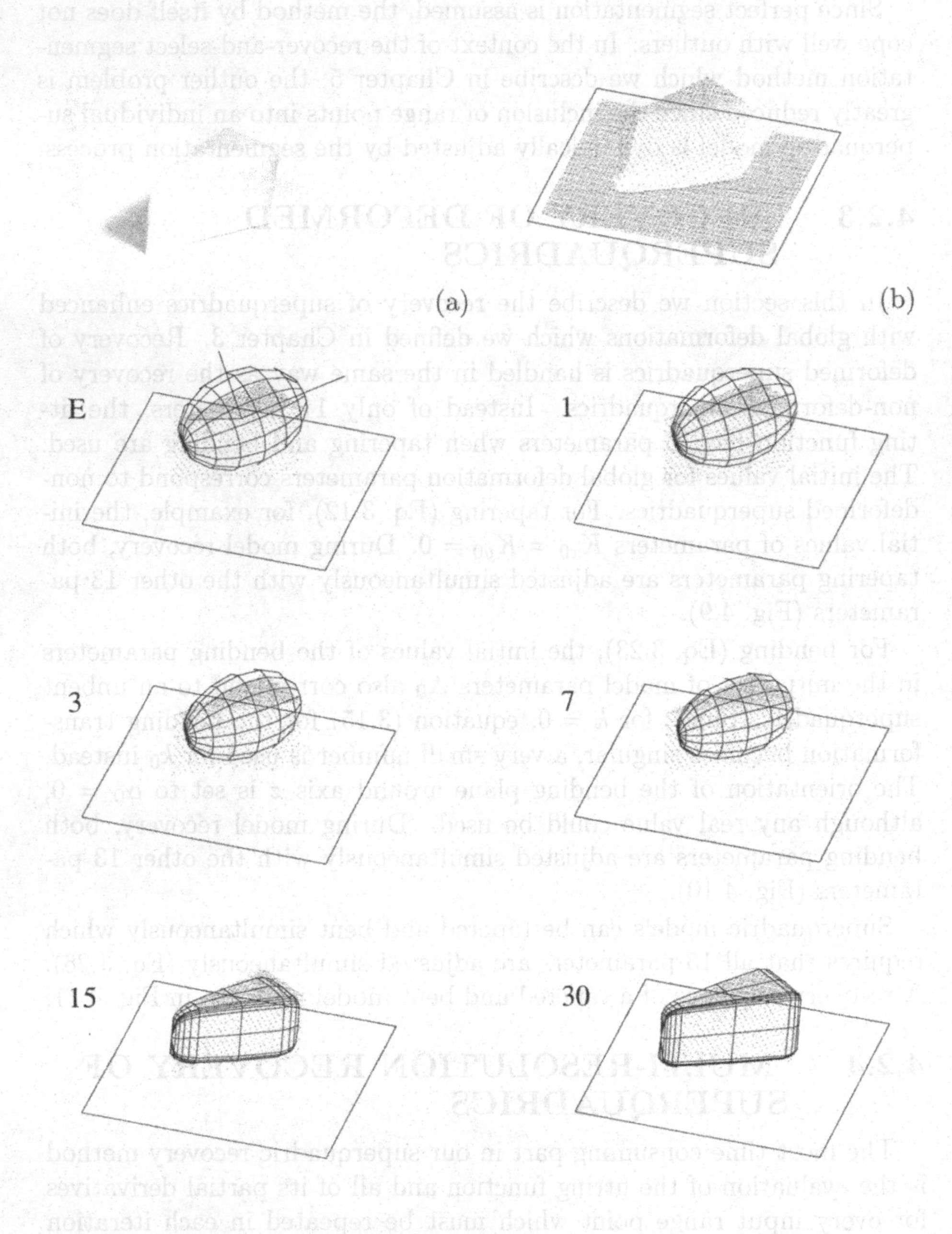

Figure 4.9. Shape recovery of a tapered object: (a) is the intensity image and (b) the corresponding range image. Below is the recovery sequence showing the initial estimate (E) and models after 1st, 3rd, 7th, 15th and 30th iteration during which all 13 model parameters were adjusted simultaneously.

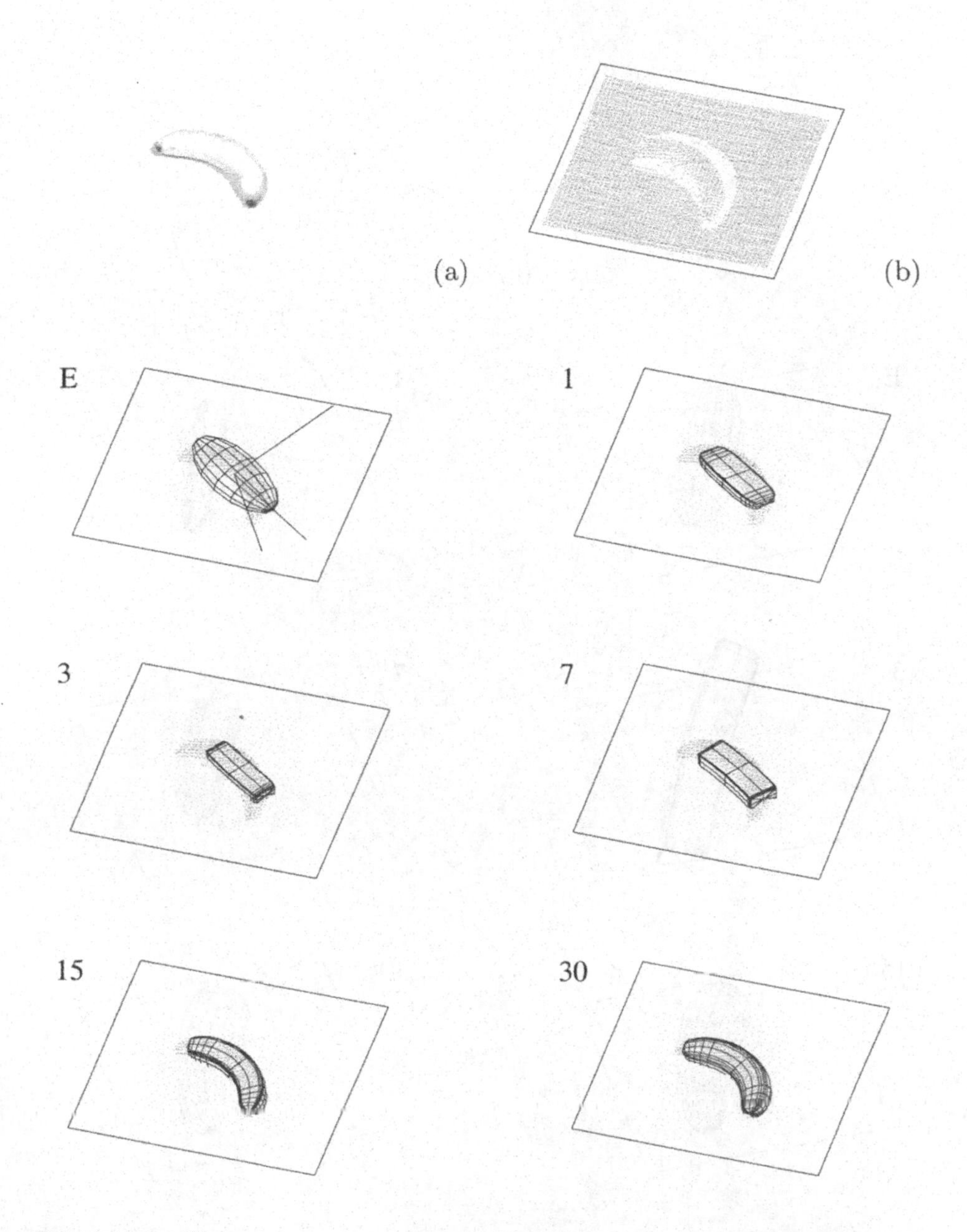

Figure 4.10. Shape recovery of a bent object (banana): (a) is the intensity image and (b) the corresponding range image. Below is the recovery sequence showing the initial estimate (E) and models after 1st, 3rd, 7th, 15th and 30th iteration during which 13 model parameters were adjusted simultaneously.

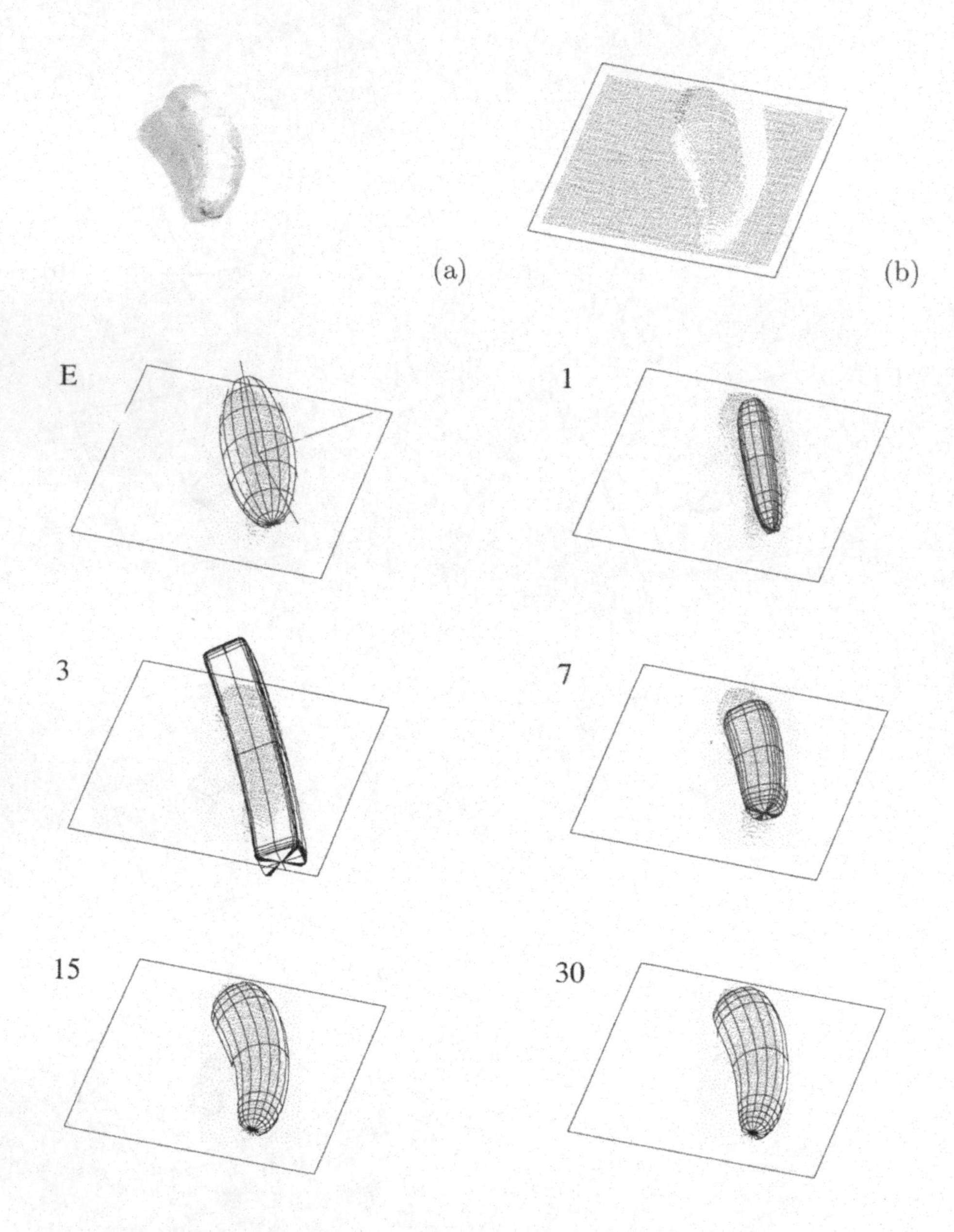

Figure 4.11. Shape recovery of a tapered and bent object (squash): (a) is the intensity image and (b) the corresponding range image. Below is the model recovery sequence when a total of 15 model parameters, including tapering and bending parameters, were adjusted simultaneously.

provement of fit (Fig. 4.5). We implemented a multi-resolution model recovery method which starts on a very coarse range image. Once no improvement in the superquadric fit is made, the minimization continues on a denser range image until the original range image is reached. In this way the minimization is up to ten times faster than without the multi-resolution scheme (Solina and Bajcsy, 1990).

4.3 PHYSICS-BASED RECONSTRUCTION

The utility of local deformations (Eq. 3.29)

$$\mathbf{x} = \mathbf{T}_{rot}(\mathbf{D}\mathbf{r}(\eta, \omega) + \mathbf{d}(\eta, \omega)) + \mathbf{p}$$

and the motion equation (Eq. 3.32)

$$\mathbf{M}\ddot{\mathbf{U}}(t) + \mathbf{C}\dot{\mathbf{U}}(t) + \mathbf{K}\mathbf{U}(t) = \mathbf{R}(t),$$

in computer vision is clear if one wishes to model nonrigid deformation of complex shapes. The external force vector $\mathbf{R}$ in equation (3.32) can be generated using image data. In turn, the shape model (Eq. 3.29) is made active under the influence of the external force vector $\mathbf{R}$. Typically, one can imagine a spring attaching every data point to the model surface. The deformed model settles in an equilibrium state with the material stiffness of the model counteracting the forces generated by the data springs. For graphics, applications such as physically-based animation (as opposed to keyframing), are obvious. The term $\mathbf{R}$ corresponds to the externally applied forces, and is most natural for animation, where the equation of motion governs the path of a body under the influence of forces.

Applications in computer vision, on the other hand, suffer from inadequate knowledge about the material stiffness ($\mathbf{K}$), the forces applied ($\mathbf{R}$), and the unknown data to model correspondence (spring attachment). This is due to the fact that "visual observation" of the original and the final state does not provide all the information needed to solve for the displacements in equation (3.32). Furthermore, the huge number of degrees of freedom ($3n$), where n is the number of nodes in the finite element model, typically makes the model recovery underconstrained.

Since the correspondence between the model and the data is, in general, unknown, the spring attachment points must be obtained by some geometric correspondence scheme, and refined during the recovery process. To ensure the topological correct correspondences, and rapid convergence, it is almost imperative to start with the model already closely aligned with the data. The spring forces and the material stiffness are relatively scaled, based on the magnitude of the desired smoothing. Furthermore, a linear elastic assumption is made to simplify the solution.

Various methods, like implicit or explicit integration, modal analysis, etc., have been used by computer vision researchers to solve the differential equation (Eq. 3.32) for unknown displacements.

4.3.1 SOLUTION OF THE MOTION EQUATION

The motion equation (Eq. 3.32) is a second-order differential equation with given boundary conditions. It can be solved (i.e., integrated) *directly* using an iterative numerical procedure. By "direct" we mean that prior to the numerical integration, the equations are used as is, without transformation into a different form. The direct integration methods achieve equilibrium at each of the discrete time points (t) within the interval of the solution. They assume a variation of displacements, velocities, and accelerations within each time interval (δt).

The simplest time integration methods (Bathe, 1982; Hildebrand, 1987) are of *explicit* type, where equilibrium conditions at time t are used to calculate the displacements at time $t + \delta t$. The methods of "central difference" and Euler integration fall into this category. The second type of time integration methods (*implicit methods*) use the equilibrium conditions at time $t + \delta t$ to determine displacements at time $t + \delta t$. The main advantage of the implicit methods is that they are *unconditionally* stable, whereas the explicit methods are constrained to follow time steps smaller than a critical time step for their numerical stability. However, the implicit methods require triangularization of the stiffness matrix $\mathbf{K}$ for the solution.

The topology of the finite element mesh determines the bandwidth of the system matrices, making $\mathbf{M}, \mathbf{C}$, and $\mathbf{K}$ of large bandwidth ($3n \times 3n$, where n is the number of nodes) and difficult to invert. In general, equation (3.32) is computationally expensive to solve by direct methods. Direct integration methods (e.g., forward Euler) are useful only when a few time steps are required (Bathe, 1982). Pentland also observed that the fitting was underconstrained due to the model having more degrees of freedom than the number of sensor observations (Pentland, 1990).

The large bandwidth of the system matrices favors techniques that either

- do not require assemblage of the entire matrices and allow element-by-element operations, or
- transform the matrices into an orthonormal modal basis, where one can reduce the system bandwidth by selecting a small number of the relevant modes and discarding the rest.

For this reason, the *explicit integration methods* and *modal analysis methods* have been most commonly used in physics-based superquadric modeling.

The explicit Euler method can be used to directly integrate the motion equation (Terzopoulos and Metaxas, 1991; Metaxas and Terzopoulos, 1993). Metaxas and Terzopoulos observed that they need not assemble the entire system matrices, but that they can define only those elements of the matrices, one by one, as they become required by the forward integration procedure. They disregard the inertial term ($\mathbf{M} = 0$) for vision applications to simplify the computation.

4.3.1.1 MODAL DYNAMICS

The method of mode superposition (Bathe, 1982) seeks to transform the equilibrium equations, such that the bandwidth of the new system matrices is much smaller, making integration more effective. The nodal point displacements $\mathbf{X}$ are transformed into *generalized* displacements $\mathbf{U}$ by

$$\mathbf{U}(t) = \Phi \mathbf{X}(t), \tag{4.33}$$

where Φ is an $n \times n$ matrix, whose columns are the eigenvectors ϕ_i (of order n) corresponding to the solution of the eigenproblem

$$\mathbf{K}\phi = \omega^2 \mathbf{M}\phi \tag{4.34}$$

obtained by postulating the solution of the motion equation (without damping) as $U_i = \phi_i \sin \omega_i (t - t_0)$.

The n eigensolutions of equation (4.34) are such that the eigenvectors are $\mathbf{M}$-orthonormal ($\Phi^T \mathbf{M} \Phi = \mathbf{I}$). The transformation in equation (4.33) gives equilibrium equations that correspond to the modal generalized displacements

$$\ddot{\mathbf{X}}(t) + \Phi^T \mathbf{C} \Phi \dot{\mathbf{X}}(t) + \Phi^T \mathbf{K} \Phi \mathbf{X}(t) = \Phi^T \mathbf{R}(t), \tag{4.35}$$

which can be written as

$$\ddot{\mathbf{X}}(t) + \tilde{\mathbf{C}} \dot{\mathbf{X}}(t) + \mathbf{\Omega}^2 \mathbf{X}(t) = \tilde{\mathbf{R}}(t), \tag{4.36}$$

where $\mathbf{\Omega}^2$ is a diagonal matrix with eigenvalues ω_i^2 on the diagonal. Thus, the finite element equilibrium equations are decoupled into n individual equations corresponding to n modes, with each mode's equilibrium equation written as

$$\ddot{x}_i(t) + \omega_i^2 x_i(t) = r_i(t) = \phi_i^T \mathbf{R}(t), \qquad i = 1, 2, \ldots, n. \tag{4.37}$$

The solution to the original motion equation is obtained by superposition of the response for each mode

$$\mathbf{U}(t) = \sum_{i=1}^{n} \phi_i x_i(t). \tag{4.38}$$

Pentland et al. used the method of mode superposition to reduce the bandwidth of the **M**, **C**, and **K** matrices (Pentland and Williams, 1989a; Pentland, 1990; Pentland, 1989a; Pentland and Sclaroff, 1991; Essa et al., 1992; Essa et al., 1993). Pentland named this approach *modal dynamics* where the general technique can be used in a variety of ways, including the static solution using mode superposition, where a closed-form solution for **U** can be achieved (after neglecting high order modes).

4.3.2 MODAL REPRESENTATIONS AND ANALYSIS

The modal analysis for description of nonrigid dynamics in computer graphics was introduced by (Pentland and Williams, 1989a; Pentland and Williams, 1989b). The discrete superquadric-inspired shape model is composed of 3D finite elements with linear interpolation functions. The dynamics of the object is simulated by solving the decoupled modal dynamics equations (Eq. 4.37) using a closed form solution. As high frequencies have no significant contribution to the object shape, speedup is obtained by discarding higher frequencies. It also has the effect of allowing larger time steps and reducing temporal aliasing. The geometric state of the model is defined by describing each mode by an appropriate polynomial function, and then constructing a global deformation matrix (corresponding to the function f in $\mathbf{s} = f(\mathbf{s}')$ in equation 3.28) to establish a correspondence between dynamic state and the geometric state. Thus, the geometric and dynamic state of the object are decoupled. Further, efficient collision detection ($O(n)$, where n is the number of points) is performed by using the inside-outside function of superquadrics. This modal representation is especially suitable for animation in computer graphics.

Pentland applied the same model and concepts as in (Pentland and Williams, 1989a) to fit superquadrics to segmented range data (Pentland, 1990). Here, instead of collision detection, the problem transforms to that of residual or force computation for fitting 3D models. Since the force term **R** is computed from image data in vision problems, the solution to the motion equation requires model to data correspondence. The correspondence is computed either along the radial line for a sphere or along the viewing direction for the superquadric model. After segmenting image data into parts by matched filtering, superquadric models are fitted by directly computing the modal deformations, either by using a gradient descent scheme, or by solving the modal dynamics equation by direct integration.

Pentland and Sclaroff followed the modal dynamics scheme to find the closed form solution to the "modal" static problem ($\tilde{\mathbf{K}}\tilde{\mathbf{U}} = \tilde{\mathbf{R}}$) in modal coordinates (Pentland and Sclaroff, 1991). However, the static solution is sensitive to model-data correspondence (also called spring attachment problem), since $\tilde{\mathbf{R}}$ is only an approximation, both in direction and magnitude. Pentland and Sclaroff refined the attachment points by repeatedly computing the solution and obtaining new attachment points. Their shape model uses the first six modes for translation and rotation, resulting in a slightly modified form of equation (3.29). Again, the high-frequency modes are ignored so as to speedup the process and to make the system overconstrained. An application of the modal technique is presented for object recognition tasks, since modes give a canonical description of the object. A reduced-basis modal description is shown to be useful for translation, rotation, and scale invariant matching. Experimental results on matching range data of a head in different orientations showed 92.5% accuracy in recognition.

The global modal deformation matrix approach developed by (Pentland and Williams, 1989a) was extended with a generalization of the superquadric explicit function by multiplying the modal deformation matrix $\mathbf{D}_u$ (whose entries are the polynomials in modes) to the displaced superquadric surface vector (Sclaroff and Pentland, 1991)

$$\mathbf{x} = \mathbf{R}\mathbf{D}_u(\mathbf{s}(\eta, \omega) + d\mathbf{n}(\eta, \omega)) + \mathbf{p} \,. \qquad (4.39)$$

This model differs from equation (3.29) in that the deformation matrix is applied to the displaced surface and that the nodal displacement is allowed only in direction perpendicular to the surface. The implicit form of the superquadric equation (i.e., the inside-outside function) derived from equation (4.39) is used for collision detection and physical simulation in computer graphics, and 3D model fitting to segmented data points in computer vision. Although the model shape employs the free vibration modes in the matrix $\mathbf{D}_u$, the model recovery is not physics-based. The $\mathbf{D}_u$ matrix is recovered by the Newton-Raphson method. The remaining surface details are described by projecting data points along the superquadric radial direction and smoothed by distributing their effect across neighboring elements by Gaussian weighting.

In computer vision, we know the deformed shape, but not the position of the control points which define the deformed shape. Further, a correspondence between the deformed and non-deformed shapes needs to be established to determine the deformed superquadric surface $\mathbf{X}$. Therefore, a fairly close initial correspondence between the data and the model is necessary. For the global alignment between the data and the model, our previously described method of superquadric recovery was used to compute the best global model for the segmented 3D point data (Bardinet et al., 1994). Then they computed the unknown positions of control points **P**, which define the local deformation, using singular value decomposition. For experiments they used synthetic data and medical images of the myocardium. Further discussion of the model-data correspondence problem can be found elsewhere (Sclaroff and Pentland, 1995; Brechbühler et al., 1995; Gupta and Liang, 1993).

4.3.3 RECOVERY OF HYPERQUADRICS

Goldgof and his research group were the first who studied the problem of hyperquadric recovery from images (Han et al., 1993; Kumar and Goldgof, 1994). Like superquadrics, hyperquadrics also do not possess a closed-form error-of-fit function based on true Euclidean distance. Goldgof used an approximation of the Euclidean distance (Taubin, 1991). For minimization, he used an iterative algorithm. Initially, just a few first terms of the hyperquadric model are recovered. Additional terms are added one by one, followed by further minimization to improve the fit to the range data. Several issues in recovery of hyperquadrics still remain open.

4.4 DISCUSSION

A direct comparison of different methods of superquadric recovery is difficult, especially only on the basis of published results. Another difficulty is that researchers have taken widely different paths to superquadric recovery, so that the methods differ in the type of input data, the definition of the cost function, as well as in the minimization methods.

For evaluation of different superquadric recovery methods, it is important to clearly define what the essential goal of a particular recovery method is. One can seek a generic part shape description of a scene as an intermediate step towards structural object recognition or, on the other hand, to get as accurate geometric models of isolated objects as possible so that objects of small geometric and size differences can be distinguished.

Input data. Much of the work on superquadrics has concentrated on the recovery from data obtained from a single view. The general problem in such a framework can be stated as obtaining a superquadric description of "sensed" data of a scene for the purpose of object classification, segmentation, recognition, navigation, manipulation, modeling, or tracking.

The sensor can be an intensity camera, a rangefinder, or a host of medical scanners like tomographic (X-ray, nuclear magnetic resonance), X-ray projection, ultrasound, etc. Each sensor acquires different modalities which can be used in a variety of ways.

Intensity sensors can be used to obtain a single image, a stereo pair of images, or multiple images of a scene, and the signal itself could be color, gray scale, or infrared. Recovery of superquadrics from 2D data is difficult, due to the loss of explicit depth information during perspective projection. As a result, the projective 2D data, in form of a color or gray image, presents a challenging task of fitting a 3-D model from a single view. Stereo data or multiple images can greatly simplify the task by providing sparse depth information which can be interpolated by the symmetry and smoothness assumptions of the 3-D model.

Range sensors can be sophisticated time-of-flight laser radars or simpler structured light systems. 3D data can be obtained from single views or by merging data sets obtained from multiple views. To consider the complexity of recovering superquadric descriptions from range images, imagine a scene consisting of a uniformly colored bright cylinder on a flat background which is imaged by a rangefinder. In the simplest case, almost uniformly sampled 3-D points on the entire surface are available as data (the background or supporting data points can be easily removed). The problem is overconstrained and there is only one sensible interpretation of the data. In such a case, no segmentation is necessary and a superquadric model can be fitted to such 3D points sets using one of several methods mentioned in this chapter. The results of most recovery methods would be, in such an ideal case, the same.

Depending on the imaging modality, however, a practical vision system is confronted with issues like noisy data, occlusions, degenerate views, incomplete data, shadows, specularities, and surface details such as text, patterns and texture. In case of range data obtained with a structured light range scanner, there might be missing range points because the structured light pattern does not reflect evenly from textured surfaces (Skočaj and Leonardis, 2000)[2]. The density of the range points further depends on the angle of incidence of the structured light. Curved surfaces are therefore unevenly sampled, and different planes may have different sampling rates, depending on the angle of the light incidence. Occlusions further contribute to uneven distribution of range points on the surface of the imaged object. Clearly, the shape recovery should be as insensitive to such biased data as possible. Objects which we wish to

[2]See Appendix C.

model with superquadrics are also not perfect geometric bodies which correspond exactly to superquadric shapes. Different superquadric recovery methods distinguish themselves primarily by how they cope with such biased data, noise, occlusions, and imperfect geometric shapes. Depending on the vision task, one can favor either generic compact models acquired by our volume minimizing method (Solina and Bajcsy, 1990) or more conservative envelopes of confidence (Whaite and Ferrie, 1991).

Objective function. Some experimental comparisons of different cost functions or error-of-fit measures for superquadric recovery from range images were done for the radial Euclidean distance versus algebraic distance (Gross and Boult, 1988; van Dop and Regtien, 1998). Van Dop and Regtien compared superquadrics recovered from several pre-segmented, noisy range images. They got somewhat better results with the radial Euclidean distance, but only after adding a background constraint and a robust technique for elimination of outliers. Cost functions for superquadric recovery are also discussed in (Gupta et al., 1989a; Whaite and Ferrie, 1991). The general problem of recovering algebraic models was studied by several authors (Taubin, 1991; Taubin et al., 1994; Sullivan et al., 1994; Keren et al., 1994; Tasdizen et al., 1999). The main drawback of representing shapes with algebraic curves seems to be the lack of repeatability in fitting algebraic curves to data. By restricting the representation to well-behaved subsets of polynomials, sufficient stability of fitting can be achieved (Tasdizen et al., 1999). A good general overview of 3D surface reconstruction methods was done by Bolle (Bolle and Vemuri, 1991).

Selecting an appropriate fitting cost function for superquadric recovery is difficult because we need to combine several, sometimes conflicting requirements. For example, the cost function should also include the uncertainty of range measurements into the recovery process. In this regard, the squared distance between the range data and the projected volume's visible surface along the depth axis z (Pentland, 1990) is a better objective function than Euclidean or algebraic distance because range data, and stereo data in particular, is much more accurate in estimating the position (x, y) than in estimating the depth (z). For this objective function, unfortunately, the gradient can be computed only by numerical means. Other, more accurate cost functions can be based on the radial Euclidean distance and even on true Euclidean distance, but these are also more difficult to implement, in particular since the true Euclidean distance cannot be computed analiticaly.

Evaluation. An important aspect of parametric model fitting is model evaluation or model validation. Model validation is essential regardless of the technique used for model recovery. The residual or the value of cost function of the final model represents an average "goodness-of-fit" over the whole model. However, some parts of the model might fit the data much better than the rest. From the superquadric model itself, it is also not evident which sections of the model are actually fitted to the image data and which sections are inferred through the parametric imposed similarity of the observed and occluded sides of the reconstructed object (Gupta et al., 1989a; Whaite and Ferrie, 1991). By explicitly representing the uncertainty of the model's surface to represent the actual surface of the object, image acquisition for improving the accuracy of representation can be planned (Whaite and Ferrie, 1993). In their work, Whaite and Ferrie described a decision theoretic framework for evaluation of recovered models which extends their earlier work on superquadric recovery (Whaite and Ferrie, 1991). This technique is embedded under the umbrella of "active" or "autonomous exploration", which evaluates the recovered model as more data is collected (Whaite and Ferrie, 1997).

Global quantitative measures like χ-square error are insufficient in pointing out specific deficiencies in model's fidelity to data. An acceptable error-of-fit also does not always result in a perceptually acceptable model, in particular if the data is incomplete and does not constrain enough the superquadric shape as in the case of 2D contours. To this end, one can study the residual of shape model recovery by comparing them against the given data. Global and local distributions of residual can be used for the objective evaluation of the recovered superquadric models (Gupta et al., 1989a; Gupta and Bajcsy, 1993). The global residual analysis provides an objective tool for determining the shortcomings of the model at the appropriate part of the object, and in determining the overall suitability of a model for given data.

Noise. Realistic conditions demand that the parametric shape models must be recovered on partial and noisy data. Noise in 3D measurements is inevitable and difficult to model. Noise occurs in the form of outlier range points mostly on edges which delineate areas of different depth where the structured light rays or laser beams split, and at the same time, illuminate both sides of the edge. Range data could be missing due to occlusions, due to the shadows in scanner geometry, or due to self-occlusion in single viewpoint data. While the symmetry constraints of superquadrics are useful in predicting the missing information, the downside of superquadric models is the lack of uniqueness in describing incomplete and noisy data within an acceptable error of tolerance.

Some of the fitting problems, in particular with outliers, which are almost always present when a pre-segmented, constant data set is given, disappear if models can grow and find their own area of applicability. Different robust techniques for elimination of outliers are based on dynamic exclusion and inclusion of range points in the current superquadric model during minimization (van Dop and Regtien, 1998). A more dynamic adaptation of the data point set to the recovered models underlies also our *recover-and-select* segmentation method (Leonardis et al., 1990; Leonardis, 1993; Leonardis et al., 1995). Chapter 5 describes how we integrated the recovery of superquadrics with this segmentation scheme (Leonardis et al., 1997). Segmentation within the framework of superquadric recovery is also discussed in (Pentland, 1990; Darrell et al., 1990).

Criteria for evaluation of superquadric recovery methods. For evaluation of a superquadric recovery method several criteria can be considered:

- one can use the cost function as a quantitative indication or global measure of how good the model fits the data points,
- one can use more qualitative criteria and judge the perceptual likeness of the recovered models with the actual objects (sometimes perceptually better results are obtained with slight under or over-fitting),
- one can evaluate the stability and uniqueness of resulting models,
- one can measure the speed of superquadric recovery, and last but not least,
- the simplicity of the implementation and the possibility of its integration with other vision tasks, such as segmentation is important.

Besides a more tight integration of model recovery and segmentation, one can also try to integrate other stages of a robotic vision system. Most methods view the superquadric part-recovery stage in isolation from later decisions (tasks) of the vision system. Hager integrated the recovery of superquadrics with decision making about the graspability of the modeled object with a robotic gripper (Hager, 1994).

We believe that our method of superquadric recovery which we described in this chapter is a good compromise of the above criteria. Our method was also successfully applied or extended in several vision or robotic systems (Allen and Michelman, 1990; Gupta et al., 1989a; Gupta and Bajcsy, 1993; Ikeuchi and Hebert, 1996; Raja and Jain, 1992; Ferrie et al., 1993; Leonardis et al., 1997; Bardinet et al., 1994).

4.5 SUMMARY

Various methods of superquadric recovery exist. In the first part of the chapter we describe several methods that differ in the type of input data, solution types, objective functions, and minimization methods.

The second part of the chapter gives a detailed description of our own least-squares gradient minimization technique for recovery of superquadrics from pre-segmented range images. We extended the recovery method also to recovery of globally deformed superquadrics. Deformation parameters for tapering and bending are recovered alongside the position, orientation, size, and the two shape parameters. We use a cost function which is based on the superquadric inside-outside function, and additionally weighted by the volume of the superquadric.

The third part of the chapter gives an overview of physics-based recovery methods. These methods are geared towards the recovery of locally deformed superquadrics and for tracking and animation.

At the end of the chapter, different methods of superquadric recovery are discussed and compared.

Chapter 5

SEGMENTATION WITH SUPERQUADRICS

A common underlying task of most recognition applications is building the scene description in terms of symbolic entities. A challenging problem in scene understanding is segmentation, where each piece of information must be mapped either to a shape primitive or discarded as noise. At the same time, there should be a minimum number of such primitives applied, so as to get as compact a description as possible. The absence of the domain knowledge further makes it more difficult, as ambiguities arise due to multiple representations and incomplete data.

There have been several attempts to segment and recover volumetric models from the data. These approaches usually involve various procedures, mostly applied in a hierarchical fashion, ranging from the estimation of local surface properties, curvature, etc., to more complex, such as symmetry seeking, in order to partition the data into parts that can supposedly be represented with a single volumetric model. Such approaches, in fact, isolate the segmentation stage from the representation stage and significant efforts are necessary to combine, for example, surface type descriptions into volumetric models.

The ability to even identify a set of surfaces as belonging to a given volume is not a trivial task without knowing at least the connectedness of surfaces, and preferably surface closure. Moreover, a surface level description may not be consistent with the volumetric description. Figure 5.1 shows an example of an L-shape object whose volumetric description can not be obtained by a simple combination of recovered surfaces. Specifically, planar patches of the surface-level description in Fig. 5.1 (c) can not be partitioned in a simple way to correspond to volumetric parts in Fig. 5.1 (d), since the top surface patch in Fig. 5.1 (c) belongs to two different volumetric models (see Figs. 5.1 (e) and (f)).

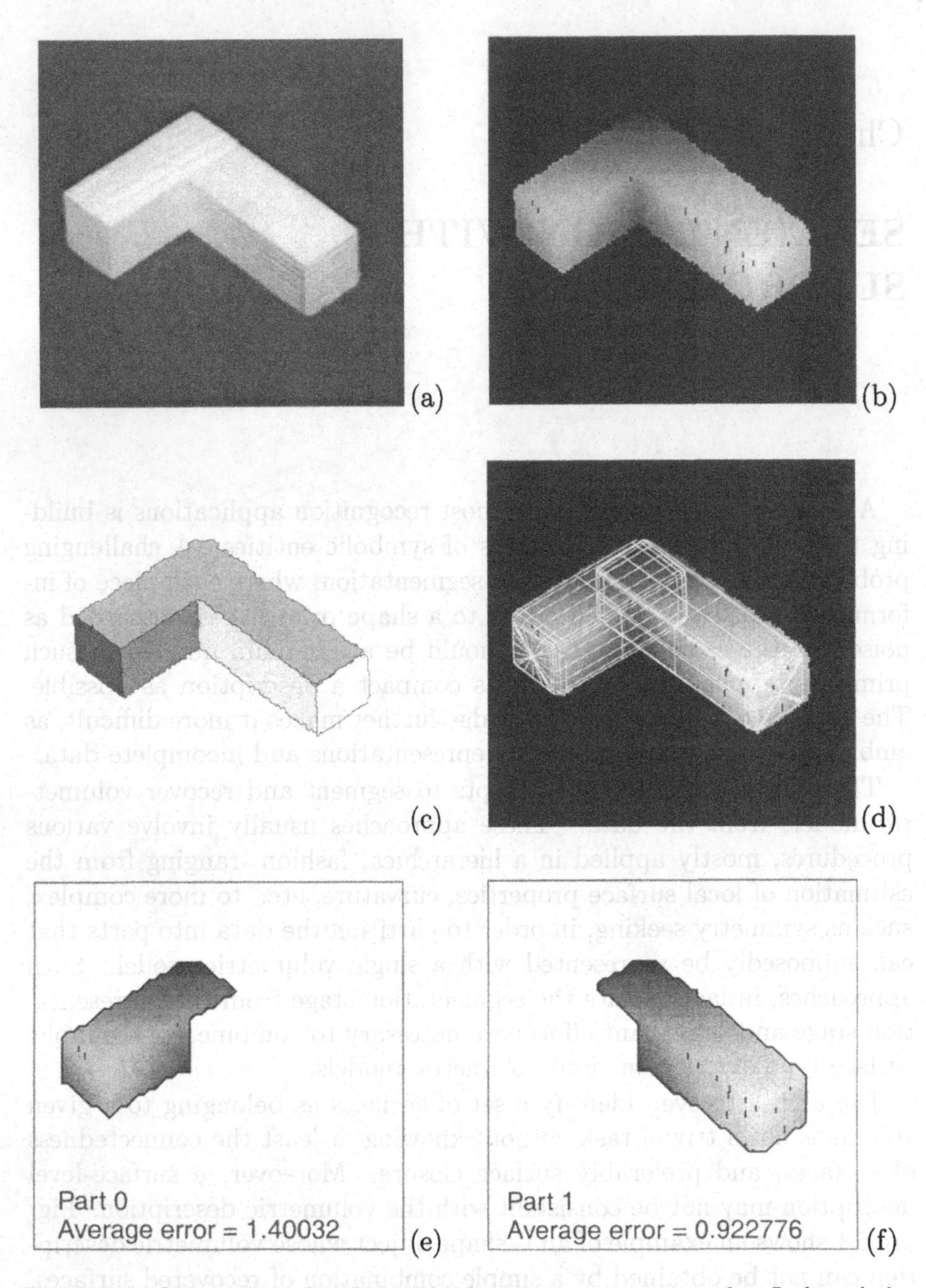

Figure 5.1. **Comparison of surface and volumetric segmentation of an L-shaped object: (a) intensity image, (b) range image, (c) planar patches of the surface-level description, (d) volumetric models, (e) and (f) domains of individual volumetric models.**

5.1 OVERVIEW OF SEGMENTATION METHODS

There have been various approaches to scene segmentation using superquadrics (Bajcsy and Solina, 1987; Pentland, 1987; Pentland, 1990; Ferrie et al., 1993; Gupta and Bajcsy, 1993; Leonardis et al., 1997).

In general, they have been applied together with the recovery of superquadric part-models from image data. The approaches to segmentation can be broadly divided into two categories:

1. **segment-then-fit** schemes: This category follows a two-stage procedure where data are segmented before and independently of model recovery. Consequently, the superquadric model is used only as a means for representation, and not for segmentation (Gupta et al., 1989b; Pentland, 1990; Ferrie et al., 1993; Darrell et al., 1990; Raja and Jain, 1994; Metaxas and Dickinson, 1993). Except for (Pentland, 1990) who uses projections of superquadrics for segmentation, methods that decouple segmentation from representation suffer from ambiguities and problems of mismatch between segmented part and the superquadric model during fitting process.

2. **segment-and-fit** schemes: This category of methods, interleaving model recovery with segmentation, is more robust (Pentland, 1987; Gupta and Bajcsy, 1993; Leonardis et al., 1997; Horikoshi and Suzuki, 1993) since the final representation also guides the segmentation. These methods follow the concept that segmentation and representation are not separable (Bajcsy et al., 1990).

5.1.1 TWO-STAGE SEGMENT-THEN-FIT METHODS

Pentland proposed a method of matched filters to segment binarized image data into superquadric part-structure (Pentland, 1990). The first part of this two stage method is based on matching 2D silhouettes (2D projections of 3D superquadric parts of different shapes, scales and of different orientations) to image data. For this segmentation stage a robust method was devised using matched filters for finding 2D silhouettes at different scales in binary images. After part segmentation, the actual superquadric recovery was performed in the second stage using a modal dynamics formulation. Here 3D superquadric models were fitted to the range data of individual part regions (range data was obtained either from laser rangefinder or depth from focus). The 3D data corresponding to each of the selected patterns was fitted with a deformable superquadric based on modal dynamics (Pentland and Williams, 1989a; Pentland and Sclaroff, 1991). The error metric used for numerical minimization is the squared distance along the z depth axis between the range data and the projected volume's visible surface. Although Pentland's segmentation method is quite robust and efficient, the assumption, of every part being visible in a 2D projection, is frequently violated in cluttered 3D scenes.

An edge-based region growing method for segmenting range images of compact objects in a pile was used by (Gupta et al., 1989b). The regions were segmented at jump boundaries, and each recovered region was considered a superquadric object. Reasoning was done about the physical support of these regions, and several possible 3D interpretations were made based on various scenarios of the object's physical support. A superquadric model was fitted and classified corresponding to each recovered object. The method was applied to sorting of postal packages, which, for the most part, can be represented by superquadrics. Problems occur at the segmentation stage, where thin overlapping objects cannot be segmented at jump boundaries.

A two-stage data-driven strategy of fitting superquadrics after segmenting range data was proposed also by (Ferrie et al., 1993). They use differential geometric properties and projected space curves modeled as snakes (energy minimizing splines). An augmented Darboux frame is computed at each point by fitting a parabolic quadric surface, which is iteratively refined by a curvature consistency algorithm. Following the transversal regularity (Hoffman and Richards, 1985) which was defined as concave discontinuity of tangent planes, the part boundaries exist at critical points defined as negative minima of the principle curvature lines. The 2D projected snakes were then used to trace the critical points. Assuming that the objects are composed of convex volumetric primitives and do not contain holes of any kind, and that the observer is free to choose its viewing position, objects can be partitioned at concave discontinuities to give parts that can be described by superquadrics. Our method for superquadric models recovery is used (Solina and Bajcsy, 1990). The authors have presented results on complex objects. Their method can fail on noisy data, and when the first critical assumption—that the objects are piecewise convex and joined at concave boundaries—is not met.

Darrell also used superquadrics to represent the surfaces segmented in range data (Darrell et al., 1990). Surfaces, described as bi-quadrics, were segmented using an MDL criterion. Each surface was assumed to map to a superquadric model without further segmentation, an assumption made by all segment-then-fit schemes. This assumption is true only when the object is piecewise convex, and all the constituent surfaces of the parts can be partitioned at the concave boundaries formed at part intersections. It is easy to see that this assumption is frequently violated in the real world, where different parts can have smoothly continuing surfaces at their intersections (e.g. the L-shaped surface of an L-shaped object in Fig. 5.1).

For a qualitative, part-based 3D object representation, Raja and Jain constructed a catalog of s-aspects[1] of all 12 geon shapes, and matched them with the recovered s-aspects of the parts of complex objects (Raja and Jain, 1994). They segment range data into surface patches, classify them according to curvature properties and generate a surface adjacency graph (SAG) encoding the surface aspect. The subgraphs of the SAG are then matched with the stored catalog to segment the SAG. The recovered parts are then identified by either directly testing the surface attributes or by fitting superquadrics and mapping them to geons, as explained before. This method of segmentation is more general than the one in (Ferrie et al., 1993) because it does not use transversal regularity as the partitioning condition for superquadric-describable parts. However, like any two-stage approach, it depends on an unrelated surface segmentation method, which affects the final result of volumetric segmentation. Furthermore, it shares with all the two-stage methods the assumption that each surface in the SAG belongs to only one part. This assumption will not allow a simple object like the L-shape to be segmented by these methods.

Another qualitative shape recovery method using geon theory was proposed by (Metaxas and Dickinson, 1993) to recover superquadrics on intensity data. Their integrated method uses a segmentation scheme based on 10 geons (Dickinson et al., 1992a) to provide orientation constraint and edge segments of a part as input for the physics-based global superquadric model recovery scheme (Terzopoulos and Metaxas, 1991). This two-stage method can handle both, orthographic and perspective projections, and also resolves depth ambiguity if stereo data is provided. But it suffers from all the above mentioned drawbacks of two-stage methods.

5.1.2 INTERLEAVED SEGMENT-AND-FIT METHODS

Recovery of superquadric models is sensitive to missing data, noise, and incorrect estimates of orientation. More precisely, model recovery

[1]surface-aspect: a set of all possible views, for which the surface types, adjacencies, and other attributes remain the same.

on partial data, which recovers a model on the predefined domain (set of points belonging to the model), will be uncertain about the shape, size, and orientation of the model. Consequently, a region growing type of method has to begin with a number of hypotheses about orientation, shape, size, and domain of the superquadric model. Since a superquadric model encodes the overall size of the domain (as opposed to the surface models like bi-quadrics), it does not lend itself to easy extrapolation (growing). Clearly, segmentation into part models must recover parts by hypothesizing parts and testing (evaluating) them.

Pentland was the first who modeled multipart range data with superquadrics (Pentland, 1987). His segmentation consists of three steps. The first step recovers superquadric models on many overlapping image regions by a coarse-grain search through the entire parameter space. This is followed by performing a gradient-descent optimization of the "best fitting" model in each image region. The best overall description of the data is obtained by picking a minimal covering of the data from among the set of regionally-best-fitting models. This method is computationally expensive due to the coarse search through the entire superquadric parameter space for each image region. The computational load can be reduced by limiting the search regions to the vicinity of the skeleton of the data, but the exhaustive search method still remains computationally expensive. Also, the skeletonization of 3-D data is by no means trivial, and is in fact similar in complexity to the segmentation problem. However, Pentland's method does demonstrate the feasibility of achieving a multi-primitive description by fitting many independent models and selecting the best among them.

In his doctoral dissertation Solina tried to segment range images of multi-part objects by running the superquadric recovery method which is described in Chapter 4 on the entire range point data set (Solina, 1987). If the initial superquadric encloses the entire set of range data points so that most of the points lie *inside* of the superquadric, the superquadric will begin to shrink once the minimization starts. By means of adaptive thresholds, points which are too far *outside* of the current superquadric are temporarily discarded from the minimization. In this way, the superquadric model shrinks until it fits to a stable data set and then grows again until it finds all the points which can be represented by a superquadric model (Fig. 5.2). Unfortunately, this simple part-segmentation method based on dynamic adjustment of the range point data set during superquadric recovery was not stable enough.

A much more sophisticated segmentation method along the same general strategy of fitting first a global model over the whole range point data set, and then splitting the model wherever the data necessitates

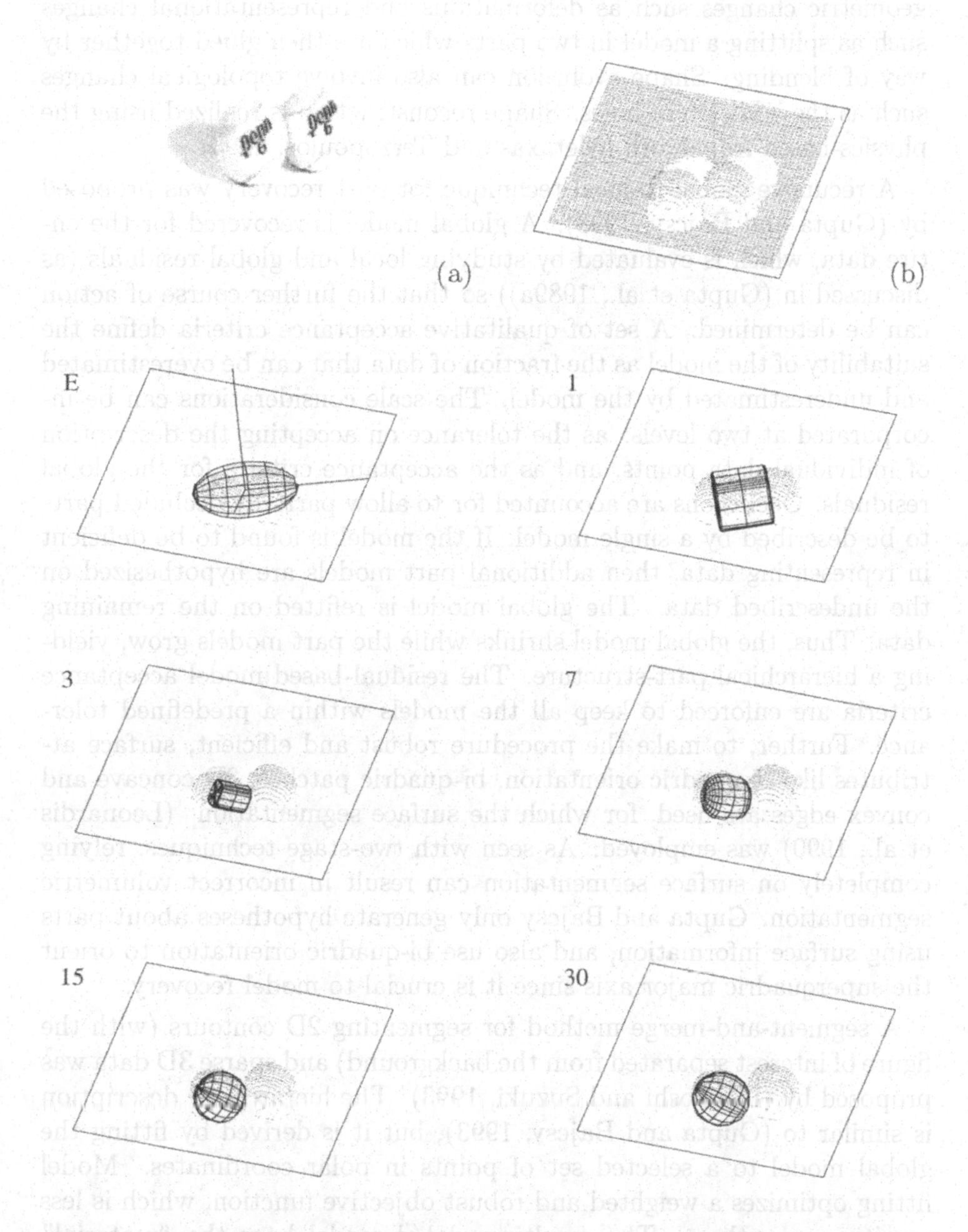

Figure 5.2. **Segmentation with superquadrics by means of model shrinking and growing (Solina, 1987). (a) is the intensity image of two tennis balls and (b) is the corresponding range image. Initial superquadric E encloses all range points, then it shrinks (iterations 1 and 3) and subsequently grows again (iterations 7, 15 and 30).**

it, was proposed by (DeCarlo and Metaxas, 1998). They named their method *shape evolution* since it is concerned with how a shape model changes during the course of fitting. The changes of the model include geometric changes such as deformations and representational changes such as splitting a model in two parts which are then glued together by way of blending. Shape evolution can also involve topological changes such as the addition of holes. Shape reconstruction is realized using the physics-based framework (Metaxas and Terzopoulos, 1993).

A recursive global-to-local technique for part recovery was proposed by (Gupta and Bajcsy, 1993). A global model is recovered for the entire data, which is evaluated by studying local and global residuals (as discussed in (Gupta et al., 1989a)) so that the further course of action can be determined. A set of qualitative acceptance criteria define the suitability of the model as the fraction of data that can be overestimated and underestimated by the model. The scale considerations can be incorporated at two levels: as the tolerance on accepting the description of individual data points, and as the acceptance criteria for the global residuals. Occlusions are accounted for to allow partially occluded parts to be described by a single model. If the model is found to be deficient in representing data, then additional part models are hypothesized on the undescribed data. The global model is refitted on the remaining data. Thus, the global model shrinks while the part models grow, yielding a hierarchical part-structure. The residual-based model acceptance criteria are enforced to keep all the models within a predefined tolerance. Further, to make the procedure robust and efficient, surface attributes like bi-quadric orientation, bi-quadric patches, 3D concave and convex edges are used, for which the surface segmentation (Leonardis et al., 1990) was employed. As seen with two-stage techniques, relying completely on surface segmentation can result in incorrect volumetric segmentation. Gupta and Bajcsy only generate hypotheses about parts using surface information, and also use bi-quadric orientation to orient the superquadric major axis since it is crucial to model recovery.

A segment-and-merge method for segmenting 2D contours (with the figure of interest separated from the background) and sparse 3D data was proposed by (Horikoshi and Suzuki, 1993). The hierarchical description is similar to (Gupta and Bajcsy, 1993), but it is derived by fitting the global model to a selected set of points in polar coordinates. Model fitting optimizes a weighted and robust objective function, which is less sensitive to outliers. The resulting global model hugs the "material" around the center of the object. This recursive procedure results in a possibly overlapping convex superquadric parts. Parts are then merged to arrive at a compact description based on Akaike's information criterion (AIC). Since merging is tried for all combinations of parts, the merging operation is exponential in computational complexity.

Our segmentation-and-fit superquadric method (Leonardis et al., 1997) uses the recover-and-select paradigm (Leonardis et al., 1990; Leonardis, 1993; Leonardis et al., 1995). This work demonstrates that superquadrics can be recovered *directly* from range data without the mediation of any other geometric models, which is in contrast to common beliefs that the recovery of volumetric models is possible only after the data has been pre-segmented using extensive pre-processing and some other geometric models. To achieve this goal, the problem of volumetric recovery was cast in the *recover-and-select* paradigm for the recovery of geometric parametric structures from image data (Leonardis et al., 1995), which was originally developed for the recovery of parametric surfaces (Leonardis et al., 1990). The paradigm works by recovering superquadric part models (fitting and classifying data points) everywhere on the image, and selecting a subset of superquadrics which give a compact description of the underlying data. This method will be described in detail in the rest of the chapter.

5.2 "RECOVER-AND-SELECT" SEGMENTATION

This section gives a thorough review of the *recover-and-select* segmentation method (Leonardis et al., 1990; Leonardis, 1993; Leonardis et al., 1995). First, we describe model recovery in the framework of recover-and-select in general and then the recovery of superquadrics as a particular model. Next, we write about model selection and different solutions to this problem. Finally, we show how model recovery and model selection can be efficiently combined.

5.2.1 MODEL RECOVERY IN GENERAL

We use the term *model-recovery* as a synonym for the process of *inductive inference* or synthesis, that is, the process in which a description is inferred from the observations (images) and *prior information.* This includes knowledge about the geometric models and also information about the sensing process because, in order to successfully accomplish the task, we need to take into account the characteristics of the noise process as well.

Here we concentrate on the model-recovery procedure in the case of parametric geometric models (superquadrics) and discuss the problems that pertain to this specific choice of prototypes. Our efforts are devoted primarily to the design of a procedure that is relatively insensitive to the kind of errors intrinsic to visual signals. First, we explain the general idea behind the design of the procedure and then elaborate on a specific algorithm that recovers superquadric models in range images.

5.2.1.1 GENERAL SCHEME

While instances of rigid models (templates) can be detected fairly easily in an image using correlation methods (Pentland and Williams, 1989a), a much more complex situation arises in the case of parametric models. There we have to find image elements that belong to a single parametric model, *and* we have to determine the values of the parameters of the model. It is evident that these two problems are tightly coupled. The phase of classifying image elements is, as any other decision making process, prone to errors (false positive as well as false negative decisions) in the presence of noise. Consequently, the parameter estimation technique has to take into account the contamination of data with measurement errors and outliers. Any simple approach to this problem is doomed to failure.

We argue that the problem of data classification and parameter estimation can successfully be solved by an iterative approach. Under the assumption that image elements have already been classified, we can determine the parameters using standard statistical estimation techniques. Conversely, knowing the parameters of the model, the search for compatible image points can be accomplished by pattern classification methods. We propose to solve this "chicken-and-egg" problem by an iterative method, conceptually similar to the ones described by (Besl, 1988) and (Chen, 1989), which simultaneously combine data classification and model fitting. Despite the similarities between our method and, in particular, the one described by Besl (Besl, 1988), our approach differs from the latter in a crucial step, namely in the selection of initial estimates (seed regions), which then leads to a variety of unique features such as redundancy, parallelism, reduced computational complexity, and provable termination of the algorithm. Besides, our method treats the recovered models only as *hypotheses* which then compete to be selected in the final description.

5.2.1.2 DATA CLASSIFICATION AND MODEL FITTING

Let us assume for a moment that we have somehow broken this vicious circle of data classification and model fitting and presume that we have a set of image elements whose statistical properties indicate that there is a high probability that they belong to a single parametric model[2]. In other words, the assumption is that the deviations of the data from the true values are mostly caused by sensor noise and the discretization error, and that the points are not contaminated by extreme measurement errors or, as they are often called, *outliers.* Later we will show that this assumption is not critical for the outcome of the final result, since our model-selection procedure has the ability to correct erroneous cases in an efficient way.

The procedure for the recovery of a single model, which can roughly be partitioned into three distinct modules, proceeds as follows (a schematic diagram of the procedure is shown in figure 5.3):

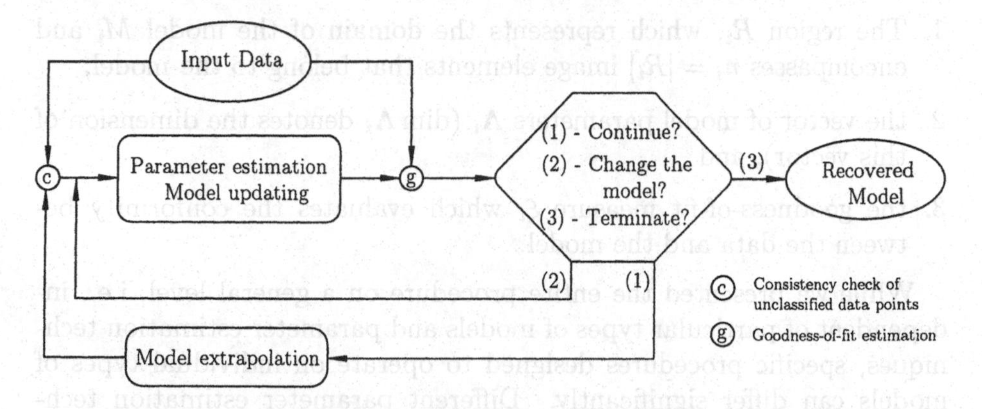

Figure 5.3. A schematic diagram outlining the model-recovery procedure

1. **Parameter Estimation:** Given a set of data points, a type of the parametric model, and an estimation technique, find an optimal set of parameters and evaluate the goodness-of-fit measure between the model and the corresponding data.

2. **Decision Making:** If sufficient similarity is established between the model and the data (goodness-of-fit), ultimately depending on the task at hand, we accept the currently estimated parameters, together with the current data set, and proceed with a search for more compatible points.

   Otherwise, a decision is made whether *to terminate the procedure*[3].

[2]We will elaborate more on this in Subsection 5.2.1.3 which deals with the problem of *seed selection.*

[3]In principle, we could at this point switch to a different type of model (e.g., higher-order models) if available.

3. **Data Point Classification:** An efficient search for more compatible points which is performed in the vicinity of the present border (endpoints) of the model is achieved through the extrapolation of the current model. Data points close to the current model are tested for consistency. Outliers, which can be classified either as extreme measurement deviations or as data points that belong to neighboring models, are rejected. New consistent image elements are temporarily included in the data set. This completes one cycle of the algorithm, and the procedure continues at step 1.

The final outcome of the model-recovery procedure for a model M_i, which has developed from the i-th seed, consists of three terms:

1. The region R_i, which represents the domain of the model M_i and encompasses $n_i = |R_i|$ image elements that belong to the model,
2. the vector of model parameters $\mathbf{\Lambda}_i$ ($\dim \mathbf{\Lambda}_i$ denotes the dimension of this vector), and
3. the goodness-of-fit measure ξ_i which evaluates the conformity between the data and the model.

While we presented the entire procedure on a general level, i.e., independent of particular types of models and parameter estimation techniques, specific procedures designed to operate on individual types of models can differ significantly. Different parameter estimation techniques are due to different dependencies of fitting functions on the set of unknown parameters. For example, a linear or a non-linear parameter estimation procedure, as in the case of superquadrics, can be a direct consequence of the choice in defining the measure of the distance between the model and the data.

5.2.1.3 SEED SELECTION

One of the crucial problems is where to find the initial regions (seeds) in an image. In our view, the strategy of selecting the seeds has a major effect on the success or failure of the overall procedure. Besides, the quality of the initial estimate influences the number of iterations required to recover a model.

It is clear that parametric models have to be initiated in a local neighborhood in order to prevent data points from different descriptions to be mixed up. Chen proposed that a window is moved around in an image, searching for an adequate amount of data that is statistically consistent in the sense that the data points belong to the same model (Chen, 1989). Thus, the requirement of classifying all points from a certain model is relaxed to finding only a small subset.

We argue that there is no guarantee that every seed will lead to a good description since the seed consistency can sometimes be satisfied on a local scale, but not on the level of a global model. As a remedy, we propose a redundant approach which can recover from a selection of globally inappropriate seeds. The idea is to independently build all possible models using all statistically consistent seeds found in a grid of windows overlaid on the image, and then use the recovered models as hypotheses that could compose the final description. The independent recovery of models has at least two important implications:

1. It prevents the erroneous initial estimates from influencing (hampering) the development of recovery of other models.

2. It offers a possibility for a fully parallelized system.

In other words, the procedure described in the previous section is performed independently, and possibly in parallel, for all seeds.

5.2.2 RECOVERY OF SUPERQUADRIC MODELS

In this section we give some details of the superquadric model-recovery in the context of the *recover-and-select* segmentation (see algorithm in Table 5.1). Procedures for seed selection, superquadric fitting, decision making regarding growing of superquadric models, and search for new compatible points (growing) are given in the following subsections.

5.2.2.1 SEED SELECTION

Initial seeds are placed on the range image in a grid-like pattern of windows. An initial seed encompasses a set of range data points in a small rectangular window whose size is determined on the basis of scale and can be adaptively changed depending on the task. A superquadric model is fitted to the data set. Next, a decision is made, whether all the data points belong to that single model. This decision, which is based on the average error-of-fit measure $\overline{\xi_i}$

$$\overline{\xi_i} = \frac{1}{|R_i|} \sum_{\mathbf{x} \in R_i} d(\mathbf{x}, M_i) = \frac{1}{|R_i|} \xi_i. \qquad (5.1)$$

compared to a threshold value *max_average_model_error* (see the next subsection on *decision making*), is not critical due to the redundant nature of the paradigm. It just eliminates those seeds that were placed on data sets that cross part boundaries and thus helps reducing the number of seeds at start of processing.

Table 5.1. Algorithm for recovery of a single superquadric within the *recover-and-select* method

```
input: a range image
    determine a seed (initial set R_i)
    fit a SQ model M_i to R_i (estimate parameters)
    if (ξ̄_i > max_average_model_error)
        the seed is rejected
    else (SQ model will develop further)
        the SQ model M_i is not fully grown
        repeat
            extrapolate/find a set of compatible points N_i
            N_i = {x : d(x, M_i) ≤ max_point_distance and
                    x ∉ R_i and x is 8-connected to R_i}
            if no new compatible points N_i = ∅
                the SQ model M_i is fully grown
            else
                fit a SQ model M_initial to data set R_i ∪ N_i
                    (using initial estimate of parameters
                    based on center of gravity and central moments)
                fit a SQ model M_previous to data set R_i ∪ N_i
                    (using parameters of M_i as 0 estimate)
                set M_temp to one of the models M_initial, M_previous
                    with minimal error-of-fit
                if (ξ_temp ≤ max_average_model_error)
                    R_i ← R_i ∪ N_i
                    M_i ← M_temp
                else
                    the SQ model M_i is fully grown
                end if
            end if
        until the SQ model M_i is not fully grown
    end if
output: recovered SQ description
```

5.2.2.2 SUPERQUADRIC FITTING

The superquadric fitting function F_S (Eq. 4.21) can be regarded as an energy function on the space of model parameters. The minimization method can, in general, only guarantee convergence to a local minimum. The starting position in the parameter space (Λ_0) determines to which minimum the minimization procedure will converge. Initial values for both shape parameters, ε_1 and ε_2 are set to 1, which means that the initial model $\mathbf{\Lambda}_0$ is always an ellipsoid. The position in world coordinates is estimated by computing the center of gravity of all range points, and the orientation is estimated by computing the central moments with respect to the center of gravity (see Sec. 4.2.2 of Chap. 4). The initial model $\mathbf{\Lambda}_0$ is oriented so that the axis z of the object centered coordinate system lies along the longest side (axis of least inertia) of the object.

To prevent sudden changes in superquadric orientation due to selection of direction of z axis for initial estimate, we use parameters of model M_i recovered from data set R_i as initial estimate for the model that is to be fitted to data set $R_i \cup N_i$. On the other hand, such a procedure might force the growing model to stay in the same local minimum as the initial model for the seed, so we also recover a model from $R_i \cup N_i$, using the initial estimate described above. We then use the model with smaller error-of-fit as a final result for the model-recovery.

5.2.2.3 DECISION MAKING

A decision whether a model should grow further or not, depends on the established similarity between the model and the data. If sufficient similarity is established, ultimately depending on the task at hand, we accept the currently estimated parameters, together with the current data set, and proceed with the search for more compatible points. The question is what could be used as an average error-of-fit measure. Due to its dependence on the superquadric size and shape parameters $(a_1, a_2, a_3, \varepsilon_1, \varepsilon_2)$, the algebraic distance (inside-outside function F, Eq. (2.16)) is not suitable. Instead, we use the radial Euclidean distance metric (Whaite and Ferrie, 1991) between a point $\mathbf{x}$ and the surface of a superquadric $\mathbf{\Lambda}$

$$d(\mathbf{x}, \mathbf{\Lambda}) = |\mathbf{x}||1 - F^{-\frac{\varepsilon_1}{2}}(\mathbf{x}, \mathbf{\Lambda})| \,, \qquad (5.2)$$

where F is given in equation (2.16). The sum over all data points belonging to the model determines the total error-of-fit ξ_i of the entire model. This is also the measure, which is, together with the number of data points encompassed by the superquadric, passed to the selection procedure.

5.2.2.4 SEARCH FOR NEW COMPATIBLE POINTS

In accordance with the paradigm, an efficient search for more compatible points is performed in the vicinity of the present border points of the region corresponding to particular model. The border points are determined on eight-connectedness and then their distance to the corresponding model is evaluated. Only those points that are close enough to the original model are included in the updated set of points. On this set of points, a new superquadric model-recovery procedure is started.

5.2.2.5 TERMINATION OF THE ALGORITHM

The algorithm always terminates, since the monotonicity requirement for growing models holds

$$\forall i: \quad R_i^{(k)} \subseteq R_i^{(k+1)} \subseteq \mathcal{I} \ , \tag{5.3}$$

where R_i denotes the domain of the i-th model in k-th growing iteration and the number of elements in the image $\mathcal{I}$ is finite.

Having in mind the unreliable initial estimates, one could think that the approach, where the points that have once been accepted can never be rejected, would almost certainly lead to erroneous solutions. Besl (Besl, 1988) claims that the monotonicity requirement, which guarantees the termination of the algorithm, is not compatible with the new pixel compatibility requirement using a simple error tolerance function. Our approach, which employs a redundant search for models in an image, however, can utilize both principles without hampering the overall performance of the algorithm.

5.2.2.6 INSENSITIVITY TO OUTLIERS

The iterative *classify-and-fit* method is an efficient tool for data-driven extraction of parametric features. Its main advantage is that the performance of the fitting is constantly monitored. The procedure dynamically analyzes data consistency allowing the rejection of outliers. The role of the *max_point_distance* threshold is twofold. Firstly, it rejects gross measurement errors, and secondly, it prevents the points which originate from neighboring models to influence the estimation procedure.

This approach, of course, presents an approximate solution, and we can not claim that all the data points belonging to a model will at the end of the procedure satisfy the compatibility constraint. However, extensive experimental testing has shown that the number of data points whose deviation exceeds the compatibility constraint is negligible. Depending on the nature of the used geometric model, the final model parameters can be improved by a post-processing procedure.

5.2.3 MODEL-SELECTION

In this chapter we elaborate the *model-selection* procedure. This involves the design of an objective function which encompasses the information about the competing parametric models and the design of an optimization procedure which selects a set of models accordingly. We also describe our approach of combining the *model-recovery* and *model-selection* procedures in a manner that significantly reduces the computational complexity.

The redundant representation obtained by the *model-recovery* procedure is a direct consequence of the decision that the search for parametric models is initiated everywhere in the image. This redundancy is reflected in the fact that several of the models are completely or partially overlapped. In other words, a single data point can be represented by several models. The final description of an image has to be obtained by fusing different partial descriptions into a globally consistent representation.

One can think of each recovered model as the best possible interpretation of the data given by an agent in terms of internal models, which together with some inference principles, encompass all its available knowledge. In our particular case, the agents differ only in their initial spatial positions[4].

The problem of fusing multiple sources of information (multiple recovered models) is a complex one. There are two options for combining individual pieces of information:

1. The final representation is formed as a simple composition (sum) of selected individual descriptions (models).

2. The final representation is formed as a composition of modified individual descriptions.

In the first case, we make the assumption that the individual descriptions produced by the model-recovery procedure are good enough to be included in the final result without modifications. The task of combining different sources of information is thus reduced to a selection procedure where the recovered individual descriptions compete as hypotheses to be accepted in the final interpretation of the image. Hence, the selection procedure is performed on the level of models rather than on the level of the models' constituent elements.

In the second case, the individual descriptions have to undergo complex interactions on the level of their constituent elements before they are combined in the final description. This implies that the number of elements involved in the procedure by far outreaches the number of models.

We follow the first approach where the task can simply be formulated as:

[4]In general, of course, they can possess a variety of mutually exclusive or inclusive pieces of knowledge.

Given a set of all recovered models together with their corresponding domains $\{(M_i, R_i) : i = 1, N\}$,
select an optimal subset of models which comprise the final result.

The core of the problem is how to define optimality criteria for choosing the "best" description. Despite the fact that the notion of optimality strongly depends on the purpose, i.e., what one wants to achieve with extracting a particular type of models, there should be mechanisms that operate on a general level. These general mechanisms should reduce the number of redundant descriptions so that a smaller number of descriptions is subsequently passed on to other modules of the system for further processing. Intuitively, this reduction in the complexity of a representation coincides with a general notion of simplicity (Leclerc, 1989b).

5.2.3.1 SIMPLICITY—LIKELIHOOD

Gestalt psychology has dealt with various phenomena regarding perceptual organization, which also includes the representation of relationships among different entities in an image. Gestalt psychologists summarized their observations in a number of Gestalt principles, one of them being the *law of Prägnanz*, or the minimum principle, which states that the visual field will be organized in the *simplest* or *the most likely* possible way. More precisely (Hochberg, 1981):

Simplicity: We perceive whatever object or scene would most *simply* or economically fit the sensory pattern.

Likelihood: We perceive whatever object or scene would, under normal conditions, most *likely* fit the sensory pattern.

The principle of Prägnanz, more than any other of the Gestalt principles, emphasizes the global qualities of perception. The intuitive notion of simplicity was experimentally analyzed (Pomerantz, 1981). Subjects were presented with dot patterns and asked to indicate the organization of each pattern as they saw it. For certain stimuli, subjects were in virtually unanimous agreement about how the pattern should be organized. Of all the possible ways in which the pattern could have been organized, the particular perceived organization minimized both the total length of the path connecting the dots and the total number of straight lines in the path. Within the same paradigm, experiments were designed where the costs and benefits of two-dimensional versus three-dimensional interpretations could be assessed. Patterns that could be organized more simply under a depth interpretation were, in fact, seen in depth. Those that could not be simplified in this way were seen to lie in the picture plane (Pomerantz and Kubovy, 1981).

Of course, the simplicity principle offers only one of several possible views on perceptual organization. It seems that proposed solutions in this difficult research domain have not yet reached a consensus. Rock (Rock, 1983), for example, presented some examples that challenge the hypothesis that the preferred perceptions are based on simplicity or regularity.

An attempt to formalize the simplicity principle will be described in the next section where the intuitive notion of simplicity is described in the framework of information theory.

5.2.3.2 MINIMUM-LENGTH DESCRIPTION

In the 14th century, William of Ockham introduced a principle, known today as *Occam's razor*, which states that *entities should not be multiplied beyond necessity.* For example, if we apply this principle to two given explanations of the data, then, all other things being equal, the simpler explanation should be preferred.

The relation between probability theory and the shortest encoding was revealed by (Shannon, 1948). Further formalization of this principle in information theory led to the principle of *Minimum Description Length* (MDL).

The MDL principle was introduced for statistical inference (Rissanen, 1983). According to the MDL principle, statistical inference is performed by choosing that model from a selected class of models which describes the data with the shortest possible encoding. He proved that when prior and conditional probabilities are known, the MDL estimator is formally identical to the MAP (Maximum *a posteriori* Probability) estimator, and *vice versa*, namely, if the probability distribution is implicitly defined on a given descriptive language, the MAP principle is equivalent to the search for the MDL (for details see Appendix D).

In psychology, the research on MDL has shed new light on a dispute over two different forms of Prägnanz (Hochberg, 1981), namely simplicity and likelihood, showing that both notions can be considered as being equivalent.

The MDL principle has found its way into computer science, including computer vision (Pentland and Williams, 1989a; Leclerc, 1989a; Darrell et al., 1990; Canning, 1991; Pilu and Fisher, 1996a; Solina and Leonardis, 1998).

In conclusion, it has been shown that specifying image models is equivalent to determining their prior probabilities. However, data compression is achievable only by an encoding scheme defining models that are suitable for a particular class of data. By specifying image models, we are hypothesizing that most or all relevant visual information can be captured in them. As a consequence, there is no such thing as an absolute measure of the simplicity of descriptions. Simplicity is always a function of one's prior assumptions (Leclerc, 1989a).

5.2.3.3 CRITERION FUNCTION (OBJECTIVE FUNCTION)

We pointed out the theoretical foundations behind the MDL principle. For practical application, however, it is sufficient to know that the minimum-length interpretation is valid when we have a coding language, which does not necessarily have to be binary. Besides, even in the absence of any knowledge of probability distributions of data, the MDL approach still maintains the highly intuitive notion of performing inductive inference by minimal coding.

We use the formalized simplicity principle only to select a subset of models from all potential models that could possibly be included in the final description. Notice that the MDL principle played no role in determining the parameters of the models or in estimating the deviations from the models, and thus did not have any influence on the model-recovery procedure. For an alternative approach which uses the MDL principle as a criterion for parametric model estimation, the reader is referred to (Leclerc, 1989a).

Let us now analyze our case. Prior to the recovery of models, the description of an image can only be given in a pointwise form. After recovering a set of models, we can describe parts of the image, or possibly the whole image, in terms of a selected subset of the set of all models. Let vector $\mathbf{m}^T = [m_1, m_2, \ldots, m_M]$ denote a set of models, where m_i is a *presence-variable* having the value 1 for the presence of a model and 0 for its absence in the final description, and M is the number of all models. The length of encoding of an image L_{image} can be given as the sum of two terms

$$L_{\text{image}}(\mathbf{m}) = L_{\text{pointwise}}(\mathbf{m}) + L_{\text{models}}(\mathbf{m}) \ . \tag{5.4}$$

$L_{\text{pointwise}}(\mathbf{m})$ is the length of encoding of individual data points that are not described by any model, and $L_{\text{models}}(\mathbf{m})$ is the length of encoding of data described by the selected models. The idea is to select a subset of models that would yield the shortest length description. In other words, we should tend to maximize the *efficiency* of the description, defined as

$$E = 1 - \frac{L_{\text{image}}(\mathbf{m})}{L_{\text{pointwise}}(\mathbf{0})} \ , \tag{5.5}$$

where $L_{\text{pointwise}}(\mathbf{0})$ denotes the length of encoding of the input data in the absence of models.

Alternatively, we can define a quantity S which represents the *savings* in the length of encoding in the presence of models

$$\begin{aligned} S &= \text{length of encoding of the data in the absence of models} - \\ &\quad \text{length of encoding of the data in the presence of models} = \\ &= L_{\text{pointwise}}(\mathbf{0}) - L_{\text{image}}(\mathbf{m}) \ . \end{aligned} \tag{5.6}$$

A similar definition was also proposed by Fua and Hanson (Fua and Hanson, 1989).

The question is how to translate the above equations into our particular case using the outcome of the model-recovery procedure. Remember that the output of the model-recovery procedure for the i-th model (M_i, R_i) consists of three terms:

1. The region R_i, which represents the domain of the model and encompasses $|R_i|$ image elements that belong to the model,
2. the vector of model parameters $\mathbf{\Lambda}_i$ ($\dim \mathbf{\Lambda}_i$ denotes the dimension of this vector), and
3. the goodness-of-fit measure ξ_i which evaluates the conformity between the data and the model.

Analogous to equation (5.4) we can write

$$L_{\text{image}}(\mathbf{m}) = \mathrm{K}_1(n_{\text{all}} - n(\mathbf{m})) + \mathrm{K}_2\xi(\mathbf{m}) + \mathrm{K}_3 N(\mathbf{m}) \ , \tag{5.7}$$

where n_{all} denotes the number of all data points in the input and $n(\mathbf{m})$ the number of data points that are explained by the selected models. $N(\mathbf{m})$ specifies the number of parameters which are needed to describe the selected models and $\xi(\mathbf{m})$ gives the deviation between the models and the data that these models describe. K_1, K_2, K_3 are weights which can be determined on a purely information-theoretical basis (in terms of bits), or they can be adjusted in order to express the preference for a particular type of description.

Now we can state the task as follows: Find $\hat{\mathbf{m}}$ such that

$$\hat{\mathbf{m}} = \min_{\mathbf{m}} L_{\text{image}}(\mathbf{m}) \ . \tag{5.8}$$

Since n_{all} is constant, minimization of equation (5.7) is equivalent to maximizing the expression

$$\hat{\mathbf{m}} = \max_{\mathbf{m}} F(\mathbf{m}) = \mathrm{K}_1 n(\mathbf{m}) - \mathrm{K}_2\xi(\mathbf{m}) - \mathrm{K}_3 N(\mathbf{m}) \ . \tag{5.9}$$

This equation supports our intuitive thinking that an encoding is efficient if:

- the number of data points described by a model is large,
- the deviations between the model and the data are low,
- while at the same time the number of model parameters is small.

So far, the optimization function has been discussed on a general level. More specifically, our objective function which takes into account the individual models has the following form

$$F(\mathbf{m}) = \qquad (5.10)$$

$$\begin{bmatrix} m_1 & \ldots & m_i & \ldots & m_M \end{bmatrix} \begin{bmatrix} c_{11} & \ldots & c_{1i} & \ldots & c_{1M} \\ \vdots & & \vdots & & \vdots \\ c_{i1} & \ldots & c_{ii} & \ldots & c_{iM} \\ \vdots & & \vdots & & \vdots \\ c_{M1} & \ldots & c_{Mi} & \ldots & c_{MM} \end{bmatrix} \begin{bmatrix} m_1 \\ \vdots \\ m_i \\ \vdots \\ m_M \end{bmatrix}$$

or in short,

$$F(\mathbf{m}) = \mathbf{m}^T \mathbf{Q} \mathbf{m} \ . \qquad (5.11)$$

The diagonal terms of the matrix $\mathbf{Q}$ express the cost-benefit value for a particular model M_i

$$c_{ii} = \mathrm{K}_1 |R_i| - \mathrm{K}_2 \xi_i - \mathrm{K}_3 \dim \mathbf{\Lambda}_i \ , \qquad (5.12)$$

while the off-diagonal terms handle the interaction between the overlapping models

$$c_{ij} = \frac{-\mathrm{K}_1 |R_i \cap R_j| + \mathrm{K}_2 \xi_{ij}}{2} \ . \qquad (5.13)$$

$|R_i \cap R_j|$ is the number of points that are explained by both models, and ξ_{ij} is defined as

$$\xi_{ij} = \max\Big(\sum_{\mathbf{x} \in R_i \cap R_j} d^2(\mathbf{x}, M_i), \sum_{\mathbf{x} \in R_i \cap R_j} d^2(\mathbf{x}, M_j)\Big) \ . \qquad (5.14)$$

The error terms $d^2_{M_i}$ and $d^2_{M_j}$ are calculated in the region of intersection $R_i \cap R_j$ and correspond to deviations from the i-th and j-th model, respectively.

The objective function takes into account the interaction between different models which may be completely or partially overlapped. However, like Pentland, we consider only the pairwise overlaps in the final solution (Pentland, 1990).

We would like to emphasize that, in contrast to some other approaches, the models whose domains R_i are completely contained within larger domains R_j are not a priori discarded but are passed to the selection procedure.

From the computational point of view, it is important to notice that the matrix $\mathbf{Q}$ is symmetric, and depending on the overlap of the models, it can be sparse or banded. All these properties of the matrix $\mathbf{Q}$ can be used to reduce the computations needed to calculate the value of $F(\mathbf{m})$.

5.2.3.4 SOLVING THE OPTIMIZATION PROBLEM

We have formulated the problem in such a way that its solution corresponds to the global extremum of the cost function. In other words, the problem of selecting the optimal set of models reduces to that of optimizing the cost of the objective function.

The optimization problem belongs to the class of problems known as combinatorial optimization. Since the function $F(\mathbf{m})$ is a quadratic function of variables that can only take the values 0 and 1, the problem is also known as the *Quadratic Boolean Problem.* The number of models M is the size of the problem. The solution space can be represented by the corners of an M-dimensional hypercube. Since the number of possible solutions increases exponentially with the size of the problem (2^M), it is usually not tractable to explore them exhaustively (except for a very small number of models). Thus it is necessary to reshape the task which means making approximations or/and being content with *suboptimal solutions.* The exact solution has to be sacrificed to obtain a practical one.

Before we proceed, we review various methods that were proposed for finding a "global" extremum of a class of non-linear objective functions that were previously thought to be practically insoluble.

Simulated annealing. Simulated annealing (Kirkpatrick et al., 1983) was first used in computer vision by (Geman and Geman, 1984). Simulated annealing is a technique that has attracted significant attention as being suitable for optimization problems of very large scale (for example, the traveling salesman problem, which is a typical case of *combinatorial minimization*). The method performs on a discrete decision space using Monte Carlo approach. This approach employs an effective temperature and annealing procedure based on the algorithms for simulating statistical mechanical systems. To avoid getting stuck in local extrema, a degree of randomness is introduced into the search strategy. The amount of randomness is controlled by the temperature parameter T which is initially set very high. As the system approaches its global optimum, the temperature is lowered. The relation between the change of the system's configuration (change of the solution vector) expressed through the energy measure E and the temperature parameter T is given by the so-called Boltzmann probability distribution

$$p(\Delta E) \sim e^{-\Delta E/kT} \quad . \tag{5.15}$$

The quantity k (Boltzmann's constant) is a constant which relates temperature to energy. If $\Delta E > 0$ then the change is accepted, if $\Delta E \leq 0$ it is accepted with a probability proportional to $e^{-\Delta E/kT}$. An annealing schedule has to be designed which determines how the temperature parameter T is lowered from high to low values (after how many random changes in configuration and in how large steps).

The main features of simulated annealing are (Hopfield and Tank, 1985):

- It causes many configurations to be averaged near a given one, which has the effect of smoothing the surface along which a search is being performed. This prevents the system from becoming stuck in local energy extrema.
- It gives the possibility of climbing out of a local extremum into another one if the annealing goes on long enough and the temperature is decreased very slowly.
- It is a method that can be used as a general method for solving non-linear optimization problems.

Hopfield's neural network. Hopfield and Tank designed a neural network (Hopfield and Tank, 1985; Hopfield and Tank, 1986) that solves problems of the following type (compare with equation (5.10))

$$E = -\frac{1}{2}\sum_{i=1}^{N}\sum_{j=1}^{N} T_{ij} V_i V_j + \sum_{i=1}^{N} I_i V_i \quad . \tag{5.16}$$

The space over which the energy function is minimized is the 2^N corners of the N-dimensional hypercube defined by $V_i = 0$ or $V_i = 1$. The search for a solution is performed inside a hypercube rather than at the corners, which means that during the search V_i can take values from $[0, 1]$. The problem is transformed into a set of coupled nonlinear differential equation. The energy function $E(\mathbf{V})$ is minimized by the gradient descent method. During the convergence, the network moves from states corresponding to very roughly defined solutions to states of higher refinement until the final solution is reached. The intermediate results represent the simultaneous consideration of many similar solutions. As analyzed by Hopfield, this general computational strategy will work well in optimization problems for which good solutions cluster, or in other words, where in the vicinity of the optimal solution, many similar solutions exist that are almost as good as the optimal one. Comparing Hopfield's neural network to simulated annealing, we have to emphasize that the neural network smoothes the energy surface during the search (at a constant level) but does not allow recovery from local minima in the solution space.

Continuation method. Pentland devised a method which could be described as the Hopfield-Tank neural network (Hopfield and Tank, 1985) placed in a temporally-decaying feedback loop. This new technique (Pentland and Williams, 1989a) is a type of *continuation method* where one first picks a problem related to the original problem that can be solved, and then iteratively solves a series of problems that are progressively closer to the original problem, each time using the last solution as the starting point for the next iteration. To be more specific, the solution of the problem given by the equation

$$E(\mathbf{x}) = \mathbf{x}^T \mathbf{Q} \mathbf{x} \tag{5.17}$$

would be straightforward if the matrix $\mathbf{Q}$ were positive or negative definite. This is, unfortunately, not the case in this type of problems. Thus, Pentland replaced the matrix $\mathbf{Q}$ with the matrix $\mathbf{Q}'$,

$$\mathbf{Q}' = \mathbf{Q} - k_3 \mathbf{I} \ . \tag{5.18}$$

The solution of the problem now proceeds as follows: The problem is first solved when k_3 is large enough to make the matrix $\mathbf{Q}'$ diagonally dominant, and thus negative definite. Then the problem is iteratively resolved by gradually decreasing the value of k_3 until the desired solution is obtained.

Pentland claims that this feedback technique produces an answer that is on average, substantially better than that obtained by Hopfield-Tank or relaxation methods. However, there are two major problems with this approach. Firstly, for large values of k_3, function $E(\mathbf{x})$ becomes convex, but of course with totally different extrema in comparison to the original problem. And secondly, the computation time to accomplish the task is significant, since for each value of k_3 (no specific prescription on how k_3 decays over time is given) the Hopfield-Tank network has to settle into a stable state.

Genetic algorithm. The term "genetic algorithm" is derived from the fact that the operations of such algorithms are loosely based on the mechanisms of genetic adaptation in biological systems. Genetic algorithms (Goldberg, 1989; Bhanu et al., 1989) were designed with the intention to be used for performing a search in an enormous hyperspace of parameter combinations using a collection of search points known as populations. By combining high performance members of the current population of solution vectors to produce better combinations of possible hypotheses, a genetic algorithm is able to find a solution vector which is located at approximate global maximum of the objective function in the problem space. The main characteristics of genetic algorithms are:

- Evaluation of the objective function for each individual solution vector determines its likelihood of affecting future generations of solution vector populations through reproduction.
- A reproduction operation produces new solutions by combining selected members of existing populations of solution vectors.
- Genetic operations involve mutations and crossovers at the stage of reproducing the solution populations.

A genetic algorithm performs its task globally since the population is distributed throughout the search space. Through reproduction operations, a genetic algorithm effectively converges to a final solution. The search from a population of individual search points is more efficient than exhaustive techniques and ideal for parallel architecture implementations.

The major drawback of the method lies in its computational complexity (evaluation of a large number of candidate solutions) which makes it infeasible for practical applications that require a prompt response.

Winner-takes-all principle. As an alternative to the algorithms discussed in the previous subsections, whose computational complexities are prohibitive even for a modest number of competing models, we consider a simple *greedy algorithm* which at any individual stage selects the option which is "locally optimal". In other words, the models are selected in the sequence that corresponds to the size of their contributions to the objective function. This is equivalent to applying, at each stage of the algorithm, the *winner-takes-all* principle[5]. This is a simple mechanism that can account for a number of phenomena that take place in brains (Koch and Ullman, 1984). Indeed, it turns out that in our case, for well-behaved inputs (well-behaved in the sense of being well describ-

[5] The term was introduced in (Feldman, 1982).

able by the chosen set of models), we obtain reasonable solutions by a direct application of the *greedy algorithm.*

Greedy algorithm. To simplify the greedy algorithm description and analysis we will use alternative notation $f(S)$ for the objective function $F(\mathbf{m})$

$$F(\mathbf{m}) = f(S), \qquad S = \{i; m_i = 1\}. \tag{5.19}$$

The $f(S)$ denotes the quality of overall description consisting of models in set S as a sum of elements c_{ij} of the matrix $\mathbf{Q}$

$$f(S) = \sum_{i \in S} \sum_{j \in S} c_{ij}. \tag{5.20}$$

Although the set S contains natural numbers as elements, we will refer to the elements as models, bearing in mind a natural bijection between the indices and the models.

Table 5.2. Greedy algorithm to approximately solve the Boolean problem

```
input: symmetric matrix Q of dimension N x N consisting of ele-
ments c_ij
output: set S of selected models
set function greedy(matrix Q)
S ← ∅
C ← {i; 1 ≤ i ≤ N}
while C is not empty
      find y ∈ C such that f(S ∪ {y}) = max_{x∈C} f(S ∪ {x})
      if  (f(S ∪ {y}) > f(S))
            S ← S ∪ {y}
            C ← C \ {y}
      else
            return S
end
return S
```

The greedy algorithm (Table 5.2) starts with an empty set of selected models S and a set of candidates C that contains all the models. At each iteration of the selection procedure a single model y from the set of candidates C is selected such that it maximizes the $f(S \cup \{x\})$ over all elements x of the set C

$$f(S \cup \{y\}) = \max_{x \in C} f(S \cup \{x\}) \tag{5.21}$$

and that the quality of the overall description increases by including the model y in the set S

$$f(S \cup \{y\}) > f(S). \tag{5.22}$$

The set S is then replaced with $S \cup \{y\}$ and set C with the set $C \setminus \{y\}$. The iteration proceeds until there is no model in set C satisfying conditions in equations (5.21) and (5.22) or the set C is empty.

The greedy algorithm is used as a heuristic which produces an overall optimal solution only under specific conditions. In order to gain at least an intuitive insight into when the algorithm produces satisfactory results, remember how the model-recovery procedure searches for instances of the models in the image. Models grow spatially as long as there are more compatible points left and the deviation between the data and the model is below a prespecified threshold. Depending on how the model growing stops, we can distinguish three cases:

1. A model whose error is within the tolerance stops growing because no more compatible points are found. Note that only the models that found a close match to the data will fall into this category.

2. A model stops growing because the initial or intermediate stages of model-recovery included outliers which caused the error term to exceed the tolerated deviation. This situation occurs when some seed regions result in erroneous initial estimates of the parameters of the models. The extent of these models is usually small in comparison to the models that started growing from the non-contaminated seeds.

3. A model stops growing because the error has slowly accumulated to the maximum value. An analysis of the residual would reveal a systematic error accumulation during the growing process.

When an area in the image is covered with models of the first type, possibly overlaid with models of the second type, the greedy algorithm will produce the optimal result. In the case of models of the third type, the greedy heuristic may give us a suboptimal solution in terms of $F(\mathbf{m})$ since it does not possess the look-a-head capability to optimally arrange the models in a piecewise description. During the course of experimentation, we observed a graceful degradation of the algorithm's performance as the agreement between the models and the data got weaker.

Fast greedy algorithm. Time complexity of the greedy algorithm equals

$$T(N) = \sum_{i=0}^{\min(M+1,N)} T_f(i+1)(N-i), \tag{5.23}$$

where N is the total number of models, $T_f(n)$ is time complexity of the algorithm evaluating $f(S)$ for the set S of cardinality $|S| = n$, and M is the cardinality of the set returned by the greedy algorithm. Obviously, M is limited from above by N, thus the same expression can be used to derive worst case time complexity simply by replacing M with N. A straightforward implementation of function $f(S)$ following the definition in equation (5.20) leads to $T_f \in O(n^2)$ to calculate $f(S)$. Since the worst case requires selection of all the models, the worst case time complexity of the algorithm is $O(N^4)$. However, since the same subexpressions of $f(S)$ are calculated over and over again, we can take advantage of this to achieve worst case complexity of $O(N^2)$.

Table 5.3. Fast greedy algorithm of worst case time complexity $O(N^2)$

```
input: symmetric matrix Q of dimension N x N consisting
          of elements c_ij
output: set S of selected models, variable f contains f(S) on return
set function incremental_greedy(matrix Q)
    S ← ∅
    C ← {i; 1 ≤ i ≤ N}
    f ← 0
    while   C is not empty
        find y ∈ C such that c_yy = max_{x∈C} c_xx
        if   (c_yy > 0)
            S ← S ∪ {y}
            C ← C \ {y}
            f ← f + c_yy
        for x ∈ C
            c_xx ← c_xx + 2c_xy
        else
            return S
  end
  return S
```

To find a model $x \in C$ satisfying conditions in equations (5.21) and (5.22), or to find out that there is no such model, it is sufficient to calculate the change of $f(S)$ if the model x is included into the set of models S

$$
\begin{aligned}
\Delta f(S, x) &= f(S \cup \{x\}) - f(S) \\
&= c_{xx} + \sum_{i \in S} (c_{xi} + c_{ix})
\end{aligned}
$$

$$= c_{xx} + 2\sum_{i \in S} c_{ix}. \qquad (5.24)$$

The change can be calculated in $O(|S|)$ time, which leads to $O(N^3)$ worst case time complexity for the algorithm. Further reduction of time complexity to calculate $\Delta f(S, x)$, can be achieved by observing that the set S is built incrementally and so can be $\Delta f(S, x)$. Suppose that the value of $\Delta f(S, x)$ is known and that we would like to know its relationship to $\Delta f(S \cup \{y\}, x)$. From the definition it follows directly

$$\begin{aligned} \Delta f(S \cup \{y\}, x) &= c_{xx} + 2 \sum_{i \in S \cup \{y\}} c_{ix} \\ &= c_{xx} + 2\sum_{i \in S} c_{ix} + 2c_{xy} \\ &= \Delta f(S, x) + 2c_{xy}. \end{aligned} \qquad (5.25)$$

Thus the $\Delta f(S \cup \{y\}, x)$ can be calculated in constant time assuming that the $\Delta f(S, x)$ is known.

At the beginning of iteration, the set S is empty so the $\Delta f(\emptyset, x) = c_{xx}$. After the set C is examined an element y is found. Then set C is replaced by $C \setminus \{y\}$ and the $\Delta f(S, x)$ values in an array are incremented by corresponding $2c_{xy}$ terms. Alternatively, the diagonal elements of matrix $\mathbf{Q}$ can be incremented by $2c_{xy}$ and the matrix dimension reduced by one. With this approach, the worst time complexity is now $O(N^2)$. We can also interpret the algorithm incremental calculation of $\Delta f(S, x)$, as a reduction of the problem of size N in time $2N - 1$, that is spent during the search for model y and to update the $\Delta f(S, x)$ values, to the problem of size $N - 1$. This leads to a non-homogeneous recurrence equation for time complexity

$$T(N) = 2N - 1 + T(N - 1), \qquad (5.26)$$

with the solution in $O(N^2)$.

It is readily apparent from the algorithm that the elements c_{ij} of matrix $\mathbf{Q}$ can be calculated on a need basis, leading to substantial computational time savings when the $M \ll N$, since we do not have to construct the whole matrix $\mathbf{Q}$ of size N^2.

5.2.4 PARAMETERS IN THE OBJECTIVE FUNCTION

There are three parameters (K_1, K_2, K_3) involved in the computation of the objective function (Eq. (5.10)). These parameters are weights which can be determined on a purely information-theoretical basis (in terms of bits), or they can be adjusted in order to express the preference for a particular type of description. In general, K_1 is the average number of bits which are needed to encode a pixel value in the absence of models; K_2 is related (as shown below) to the average number of bits needed to encode a residual value in the presence of models, and K_3 is the average cost of encoding a parameter of the model.

In order to determine the value of K_2, let us suppose that deviations of the data from the models are modeled as zero-mean Gaussian noise with variance σ^2. The length of encoding of a discrete independently identically distributed (IID) random process, that is, a sequence of random variables $x_1, x_2, \ldots, x_n$ with the Gaussian probability distribution $p(x) = N(0, \sigma)$, is determined as

$$\begin{aligned} L_{\text{deviations}} &= -\log_2\left(\prod_{i=1}^{n} p(x_i)\right) = \\ &= -\sum_{i=1}^{n} \log_2 p(x_i) = \\ &= -\sum_{i=1}^{n} \log_2 \left(\frac{1}{\sqrt{2\pi}\sigma} \exp\left(-\frac{x_i^2}{2\sigma^2}\right)\right) . \end{aligned} \tag{5.27}$$

More specifically,

$$L_{\text{deviations}} = -\log_2 \exp\left(-\frac{1}{2}\sum_{i=1}^{n}\frac{x_i^2}{\sigma^2}\right) + n(\log_2\sigma + \frac{1}{2}\log_2 2\pi) . \tag{5.28}$$

We estimate $\sigma^2 = \frac{1}{n-1}\sum_{i=1}^{n} x_i^2$. The first term on the right side of equation (5.28) can be expressed as

$$-\log_2 \exp\left(-\frac{1}{2\sigma^2}\sum_{i=1}^{n} x_i^2\right) = \frac{1}{2\sigma^2}\sum_{i=1}^{n} x_i^2 \log_2 e = \frac{n-1}{2}\log_2 e . \tag{5.29}$$

Equation (5.27) can now be written as

$$\begin{aligned} L_{\text{deviations}} &= \frac{n-1}{2}\log_2 e + n(\log_2\sigma + \frac{1}{2}\log_2 2\pi) \approx \\ &\approx n(\log_2\sigma + \frac{1}{2}\log_2 2\pi e). \end{aligned} \tag{5.30}$$

The second term in equation (5.12), $K_2\xi_i$ ($= K_2\chi^2$), handles the deviation between the data and the model. By comparing the term $K_2\chi^2$ to $L_{\text{deviations}}$ (Eq. (5.30)), we get an approximate value for K_2

$$K_2 \approx \frac{\log_2\sigma + \frac{1}{2}\log_2 2\pi e}{\sigma^2} . \tag{5.31}$$

Note that the value of K_2 depends on the standard deviation of the noise distribution and decreases with an increase of the noise level σ. However, when σ is known to be within a range of values, we can calculate an approximate value for K_2 and keep it constant for that particular case.

More pragmatically, we can determine suitable values for K_1, K_2, K_3 by considering a few limiting cases. Since we are interested only in the relative comparison of possible descriptions, we can set K_1, which weights the number of data points, to 1 and normalize K_2 and K_3 relative to it. If we assume for a moment that the error equals 0, K_3 determines the minimum number of data points that a model has to encompass in order to be preferred over a pointwise description. Once we have obtained the value for K_3, we can, by fixing the maximum allowable error and the minimum number of data points in the model, determine the value of K_2. A list of specific values of the parameters K_1, K_2, and K_3 are given in Chapter 6.

5.2.5 COMBINING MODEL-RECOVERY AND MODEL-SELECTION

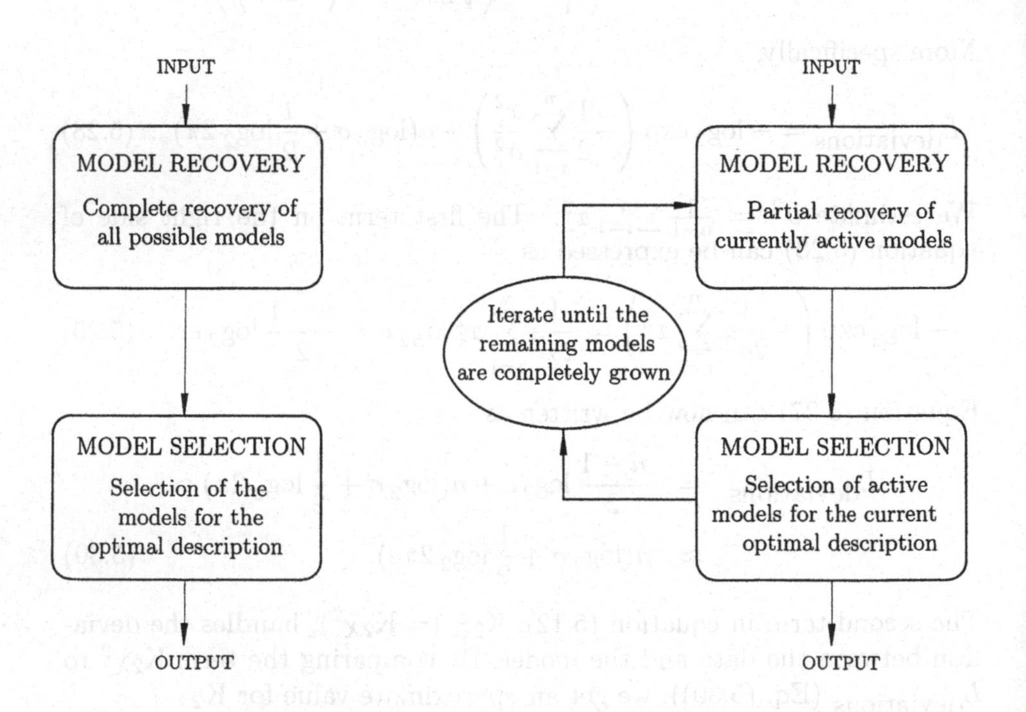

Figure 5.4. Recover-then-select strategy

Figure 5.5. Recover-and-select strategy

After having explained the two major components of our system, namely the modules for model-recovery and model-selection, we now describe how they can be combined to obtain a fast and efficient overall method. For example, the modules for model-recovery and model-selection can be applied in succession or in a loop (Fig. 5.5).

If the two modules are applied in succession, all the models are first grown to their full extent and then passed to the selection module. As a consequence of the selection process, eventually very few of the recovered models emerge as acceptable descriptions of the data. We call this procedure *Recover-then-Select* because it grows all the models fully and then discards the redundant ones. However, the computational cost of growing all the models completely is prohibitive in most cases. Instead, it would be desirable to discard some of the redundant and superfluous models even before they are fully grown. This suggests incorporating the selection procedure into the recovery procedure, as shown in Fig. 5.5. We call this approach *recover-and-select.* The inclusion of superquadric models within the recover-and-select approach is outlined in Table 5.4.

Before we proceed, we have to answer the question whether we can expect that the results obtained by the recover-and-select procedure will be comparable with the results obtained by the recover-then-select procedure. Extensive experimentation has shown that when there is a close correspondence between the models and the data, the final values of the objective function are almost identical in both cases (within a few percents). This outcome is expected since the models that initially accumulate substantial errors and whose majority of data points which they cover is better described by other models, are models that incorporated outliers and can thus be rejected very early in the process. On the other hand, the selection among the redundant (overlapped) models which are approximately of the same extent and error, does not have a major impact on the final result since all these models would yield almost the same result at the end.

However, if the models are inadequate to describe the data, we can expect that, by not waiting with the selection process till the end of the growing, the result may be a non-optimal description.

The recover-and-select paradigm opens up a number of possibilities how to control the model growing procedure. We can identify several trade-offs in combining the model-recovery module with the module for the selection. The more the models are grown, the more reliable is the description they give. But the initial growing is computationally expensive and also results in a less sparse matrix **Q**. However, this reduces further processing since fewer models are selected for further growth. On the other hand, if the growing process is interrupted by the selec-

Table 5.4. **Algorithm for superquadric recovery and segmentation within the *recover-and-select* paradigm**

```
input: a range image
determine a set of seeds (initial sets R_i)
for all seeds do
    fit a SQ model M_i to R_i(estimate parameters)
    if (ξ_i ≤ max_average_model_error) put the SQ model
        into the set of currently active, not fully grown models
    else
        the seed is rejected
    endif
end for
steps ← length of the square shaped seed region / 2
while there are any active, not fully grown SQ models do
    for i = 1 to steps do
        for all active not fully grown SQ models do
            extrapolate/find a set of compatible points N_i
            N_i = {x : d(x, M_i) ≤ max_point_distance and x ∉ R_i
                and x is 8-connected to R_i}
            if no new compatible points N_i = ∅
                the SQ model is fully grown
            else
                fit a SQ model M_initial to data set R_i ∪ N_i
                    (using initial estimate of parameters based on
                    center of gravity and central moments)
                fit a SQ model M_previous to data set R_i ∪ N_i
                    (using parameters of M_i as initial estimate)
                set M_temp to one of the models M_initial, M_previous
                    that has the minimal error-of-fit
                if (ξ_temp ≤ max_average_model_error)
                    R_i ← R_i ∪ N_i
                    M_i ← M_temp
                else
                    the SQ model M_i is fully grown
                end if
            end if
        end for all
    end for
    perform selection among all active SQ models for the current
    optimal description (only selected SQ models remain active)
    steps ← 2 * steps
end do
output: part-level (SQ) description of a range image
```

tion of currently optimal models at the early stages, the complexity of the early processing is reduced and the matrix **Q** is sparse due to less overlapping. In this case, fewer models are rejected, increasing the complexity of the further processing. These trade-offs are summarized in Table 5.5. By properly balancing the two trade-offs, a computationally efficient and reliable algorithm is obtained which has the feature of growing only well-behaved models (in terms of convergence, error, and number of compatible points), while at the same time lowering the computation time and space complexity of the procedure.

Table 5.5. Trade-offs in combining model-recovery and model-selection procedures.

Feature	Model-selection invoked	
	Early	Late
Description	– Less reliable	+ More reliable
Models remaining for further growth	– More	+ Fewer
Processing needed for initial growing	+ Less	– More
Matrix **Q**	+ Sparse	– Dense

A model-recovery procedure is interrupted by model-selection. An important decision is when to perform a selection. We choose to interrupt the recovery procedure, when at least one model reaches twice its original size (4 times its area) in the image or there are no models left that can grow any further. Since initial seeds are placed in a grid like manner, the twice the original size criterion ensures sufficient overlap of the models, so that the number of models is reduced while it is still not too computationally expensive to let them grow to such a size.

5.3 SEGMENTATION WITH SURFACE MODELS

The recover-and-select segmentation method was initially developed for surface models (Leonardis, 1993; Leonardis et al., 1995). In Figs. 5.6 and 5.7 are given two examples of segmentation using second-order surface models. Results of segmentation with superquadric models can be found in Chapter 6.

The second-order surface in Fig. 5.6 (see also (Fan, 1989)) is difficult to segment due to the absence of a jump or a surface normal discontinuity. Our method gives a clean separation of the curved surface from the neighboring planar surfaces. Such a result is possible only because we

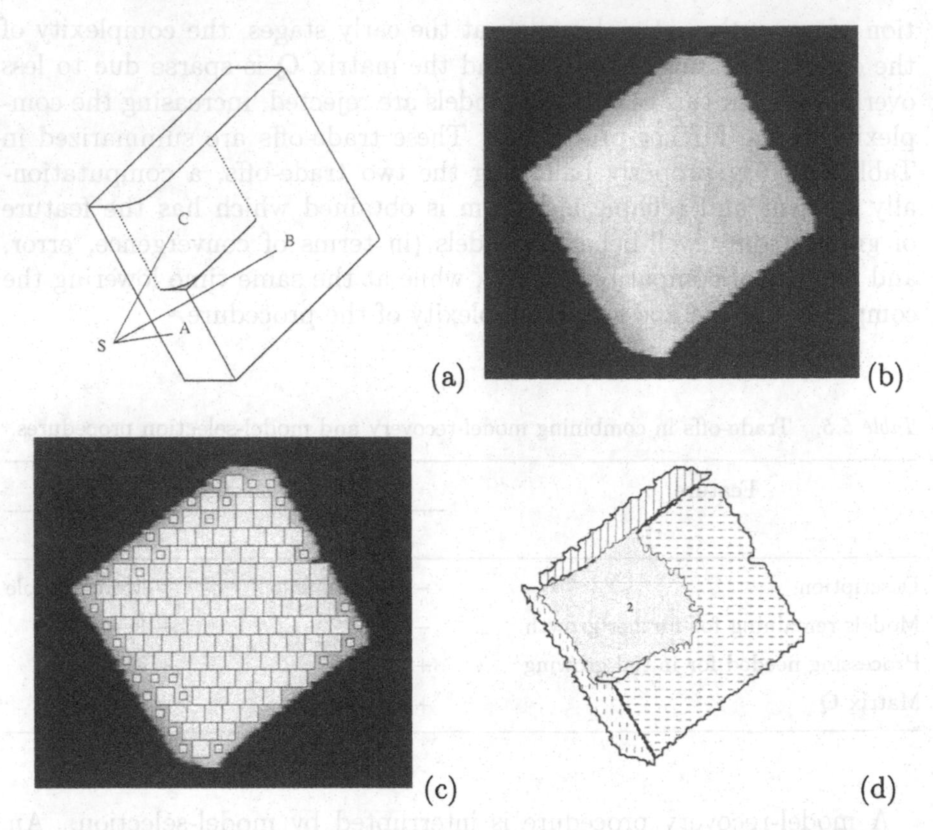

Figure 5.6. Complex object: (a) CAD model, (b) original image, (c) seed image, (d) segmented image (the second-order patch (2) is segmented from the planar patches).

search for the best description everywhere in the image and allow the models to develop independently.

The noisy range image of a cube in Fig. 5.7 (taken from (Besl, 1988)) has both planar and curved surfaces. Due to the noise in the image, a higher value was used for the compatibility constraint. This has the effect of non-uniform smoothing where the noise points are left out in the final description. The data points that were detected as outliers are marked in black (Fig. 5.7 (c)). The segmentation is as good as the one obtained by Besl using a curvature-based approach, but processed at a fraction of the computational cost involved in such elaborate processing.

5.4 SUMMARY

We present a general scheme for segmentation and recovery of parametric models which we named *recover-and-select*. The major novelty of the recover-and-select segmentation scheme is that the parametric models are simultaneously and independently recovered by growing models

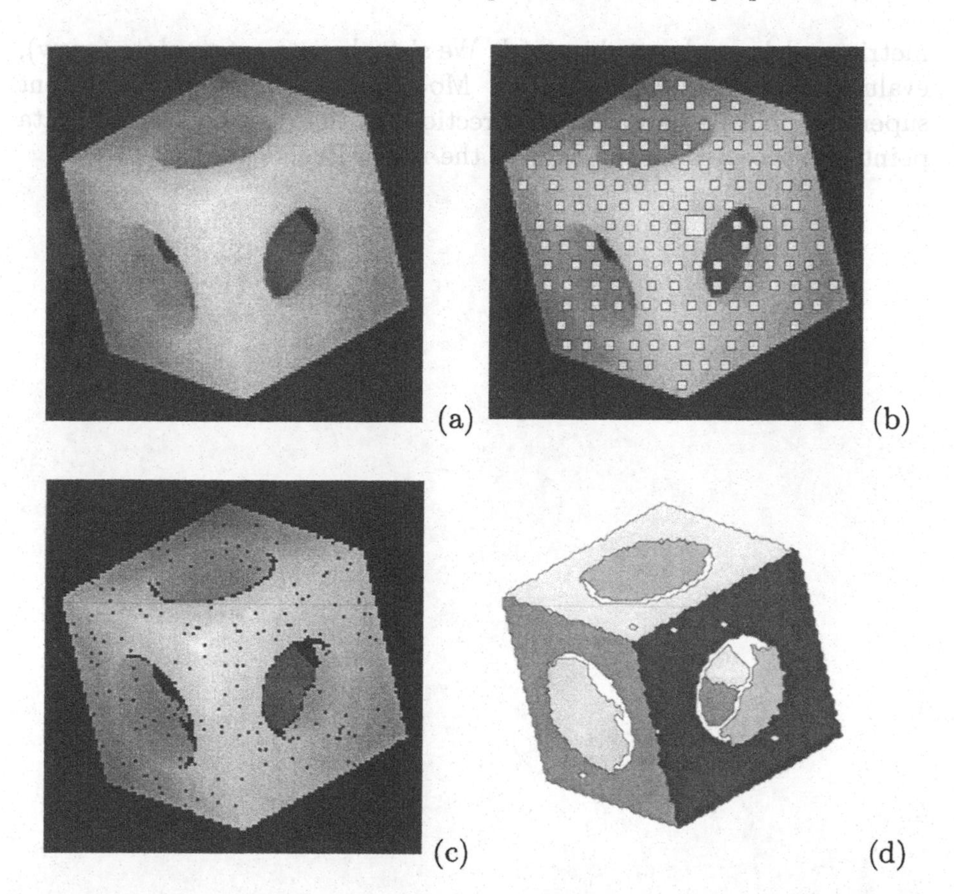

Figure 5.7. Cube: (a) original image, (b) seed image, (c) reconstructed image, and (d) segmented image

from the initial *seeds*, which results in a redundant representation. From this redundant representation, a compact MDL-like representation is obtained by selecting models for the final description which is posed as an optimization problem. Optimization of the objective function that encompasses information about the competing models is performed by a simple winner-takes-all algorithm which turns out to be a good compromise between speed and precision.

Finally, we show how to reduce the computational complexity of the overall segmentation and recovery method by interleaving the model-recovery and model-selection procedures. Since it is computationally prohibitive to grow all possible hypothesized part-models to their full extent, and then selecting the final models, best models are selected after every few iterations of model-recovery.

The recover-and-select segmentation method was originally developed for surface models. In this book we demonstrate how superquadric volumetric models can be used instead. We show how to extrapolate (grow), evaluate, and select superquadrics. Model extrapolation of the current superquadric is performed in all directions so that new compatible data points are accepted on the basis of the radial Euclidean distance.

Chapter 6

EXPERIMENTAL RESULTS

We tested the proposed segmentation algorithm on a variety of real and synthetic range images. Most of the objects in these images were composed of parts that can be modeled well by non-deformed superquadrics. In addition, we also tested the performance of the segmentation algorithm on some objects that cannot be perfectly modeled by non-deformed superquadric.

We acquired most of the range images for our experiments with a structured light range scanner (Skočaj, 1999). The scanner and the method of obtaining range images of objects are described in Appendix C. Experiments were performed with the *Segmentor*, our object-oriented framework for image segmentation (Jaklič, 1996). For details on *Segmentor* and how to obtain it, see Appendix E.

In the first part of the chapter, we present the experimental results of segmentation on several range images. In particular, we were interested in

- the quality of segmentation in terms of the number of recovered superquadrics and their qualitative similarity with the actual parts.

An analysis of our initial experiments with the basic recover-and-select paradigm revealed its intrinsic limitation of the precision of segmentation and consequently, the precision of recovered models. To improve the precision of recovered superquadrics, we developed a post-processing method. The second part of the chapter therefore consists of a more thorough analysis of the segmentation method. In particular, we tested

- the quality of segmentation in terms of errors of individual recovered models,

- the sensitivity of the segmentation to changing viewpoint,
- the sensitivity of segmentation to measurement noise and outliers present in range images, and
- the performance of segmentation on data that cannot be perfectly modeled by superquadrics.

6.1 RECOVER-AND-SELECT SUPERQUADRIC SEGMENTATION RESULTS

The segmentation method was tested on a variety of real range images. To reduce the computational burden which is mostly dictated by the numerical minimization during recovery of superquadrics, and which in turn, depends on the number of range points, the range images of size 450×450 pixels were subsampled by a factor of three. Initial size of the seeds was 12×12 pixels, which is 4×4 pixels after subsampling. The thresholds needed in the recover-and-select algorithm were experimentally determined. The actual values of these thresholds which were kept constant throughout the experiments shown in this section, are given in Table 6.1.

Table 6.1. Thresholds for model recovery and constants for model selection

max_point_distance	*max_average_model_error*	K_1	K_2	K_3
2.1	1.0	1.0	0.3	1.0

Superquadrics seeded on planar regions would become, during recovery, exceedingly thin. To prevent numerical degeneracy during the minimization of the fitting function of such superquadrics, we used a projection method to limit the size of superquadric parameters a_1, a_2, a_3 above 1.0. In most cases, 15 iterations of the Levenberg-Marquardt method were sufficient for convergence of the minimization.

On the average, it took less than 30 minutes to process each range image on an Intel Pentium II 400 MHz platform running Linux. Note that the processing time for segmentation mainly depends on the number of superquadric growth iterations to reach the final size of the models. Therefore, the segmentation of a large object which requires a larger number of initial seeds, as well as more growth iterations, takes more computing time than the segmentation of a smaller object. However, the processing time for superquadric segmentation is not critical since individual models could be recovered in parallel. Besides, the computation

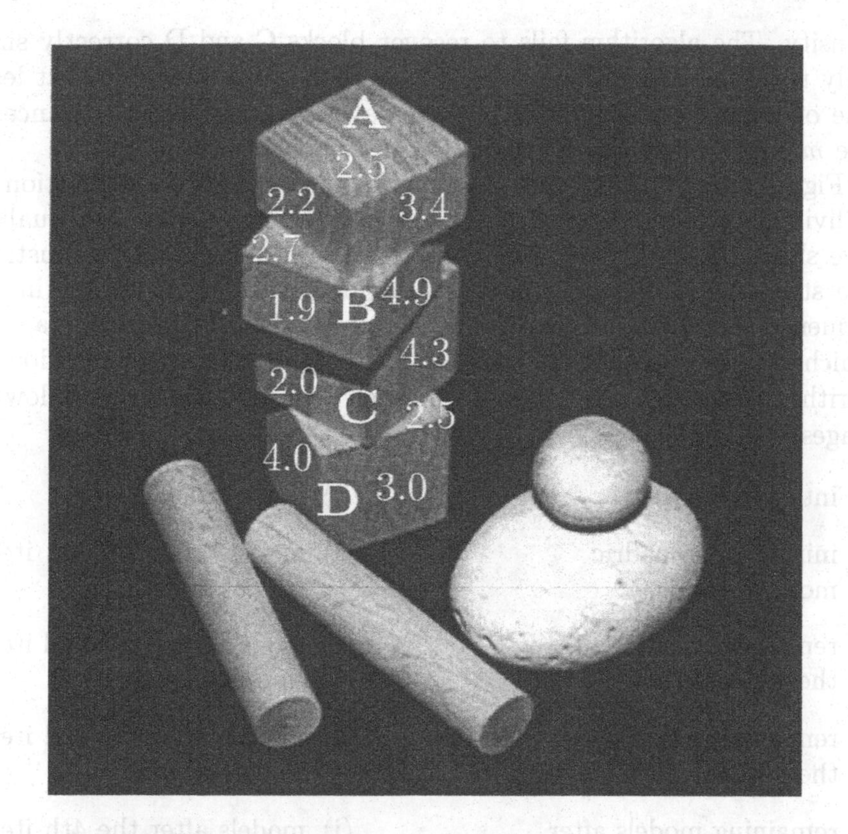

Figure 6.1. Average distance between neighboring range data points depends on the local surface orientation, relative to the sensor system. The actual average range point distances on surfaces of blocks A, B, C, and D in the range image are indicated on the corresponding intensity image shown above.

of the superquadric fitting function and its derivatives is independent for each range point and can also be parallelized at a fine grain level in a straightforward way.

The structured light range scanner that we used for our experiments produces range data whose density depends on local surface orientation. Fig. 6.1 shows how the average distance between neighboring range data points on planar surfaces of a stack of blocks depends on surface orientation, relative to the sensor system. The topmost block A has three completely visible sides, while the blocks B, C, and D below have progressively less visible sides. Blocks A and B are successfully recovered (Fig. 6.6) since two sides of each of the two blocks have average range point distances close to *max_point_distance* = 2.1. The third side, with low range data density, gets included into the region naturally, as the superquadric model grows over the two sides with the larger range point density. The algorithm fails to recover blocks C and D correctly since only two sides are visible in the subsampled range image, and at least one of them has a significantly larger average range point distance as the *max_point_distance* threshold for growth.

Figs. 6.2, 6.3, 6.4, and 6.5 demonstrate correct identification of individual superquadric parts in terms of their number and qualitative shape. Each figure presents a sequence of images which illustrate the stages in the *recover-and-select* segmentation. Each image in the sequence shows the original range image overlaid with superquadrics, which were recovered up to the particular stage of the segmentation algorithm. The individual images in each sequence show the following stages:

(a) intensity image,

(b) input range image,

(c) initial superquadric models—seeds,

(d) models after the 1st iteration of growth,

(e) remaining models after the 1st selection,

(f) models after the 2nd iteration of growth,

(g) remaining models after the 2nd selection,

(h) models after the 3rd iteration of growth,

(i) remaining models after the 3rd selection,

(j) models after the 4th iteration of growth,

(k) remaining models after the 4th selection,

(l) final models on the intensity image.

The processing of most range images in this chapter with the recover-and-select segmentation method required four iterations of growth and selection. The processing of some range images, however, required less or more than four iterations of growth and selection to reach the final result. This depends on the size of the final models. In such cases, the intermediate results are replicated or omitted, respectively.

Fig. 6.2 shows the segmentation of individual grape berries for which superquadrics are a very accommodating model. For correct segmentation, it would suffice that a single superquadric seed would be placed on each berry. Three growth and selection iterations are enough to reach the final segmentation, due to the small size of individual berries.

Fig. 6.3 shows the segmentation of a bunch of grapes where individual berries overlap with each other. Superquadric seeds were not placed on the branches because at the given resolution the selected seed size was too large, and therefore also the average error-of-fit measure $\overline{\xi_i}$ (Eq. 5.1) was too large. A single superquadric seed which was placed on the branches did not evolve, indicating thereby that these structures are too thin to be modeled with superquadrics at the selected scale.

Fig. 6.4 shows how occlusion influences the growth of superquadric models. Even if an object is almost completely occluded through the middle by another object, one superquadric model eventually grows into a complete model of the occluded object since the corresponding region is connected. If the occlusion produces unconnected regions, however, not a single superquadric model can bridge the gap. The result are two superquadrics models recovered for the left and for the right part of the occluded object.

Fig. 6.5 is an example of a complex scene which consists of eleven parts: five ellipsoids, four parallelepipeds, and two cylinders. The final result shows how touching parts can affect the final superquadric size due to incorrect classification of image points.

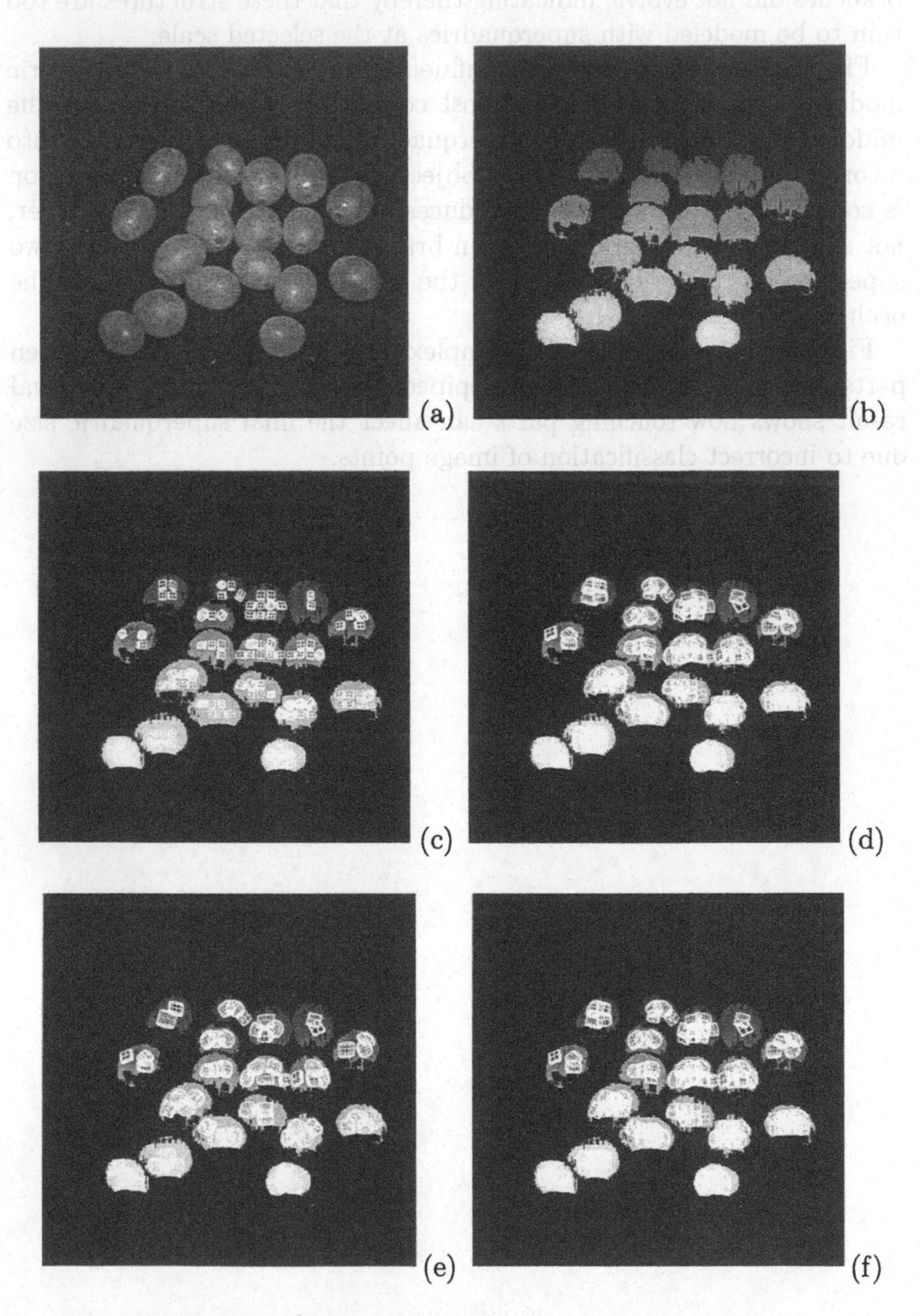

Figure 6.2. Segmentation of non-overlapping grape berries.

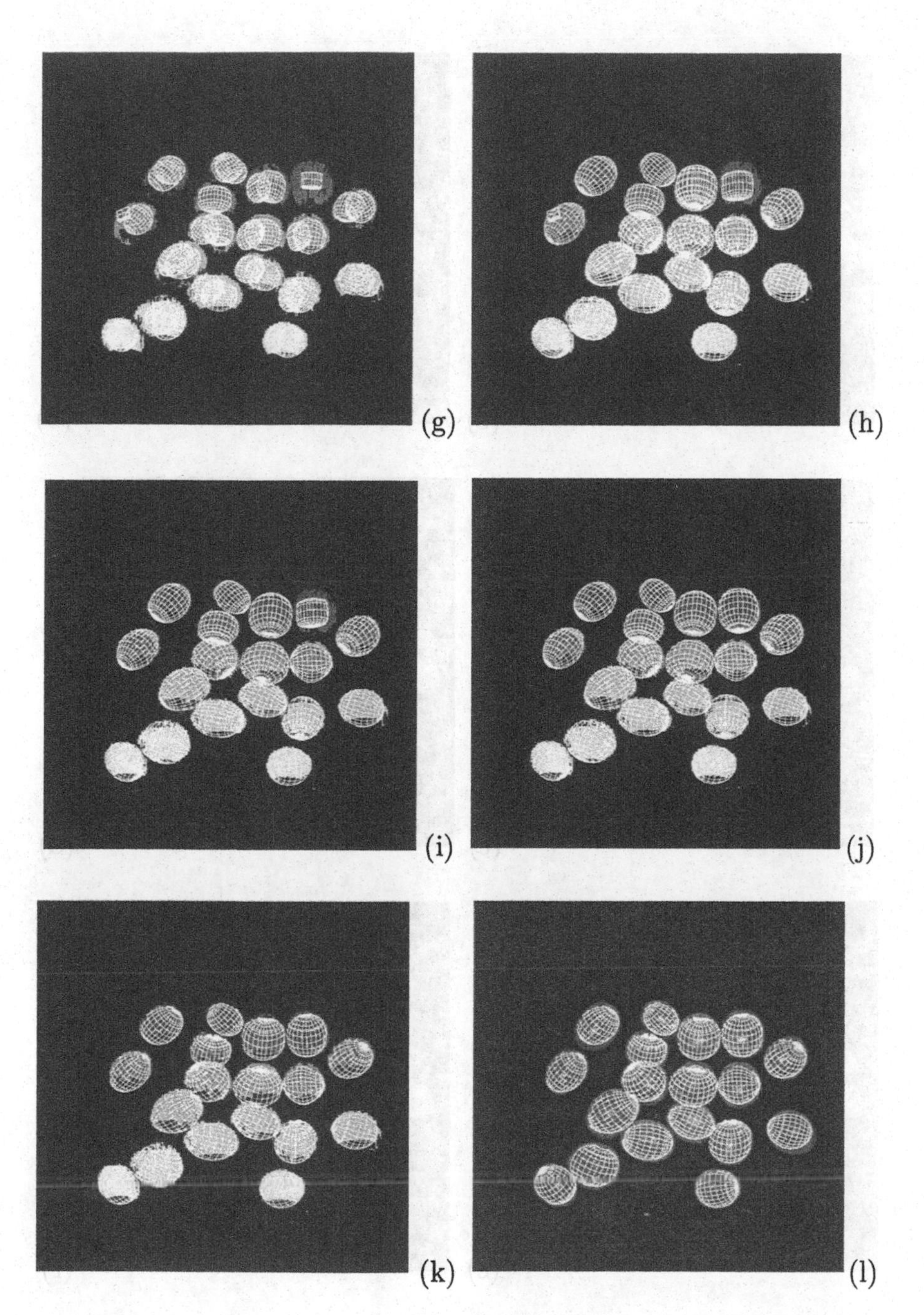

Figure 6.2 (continued).

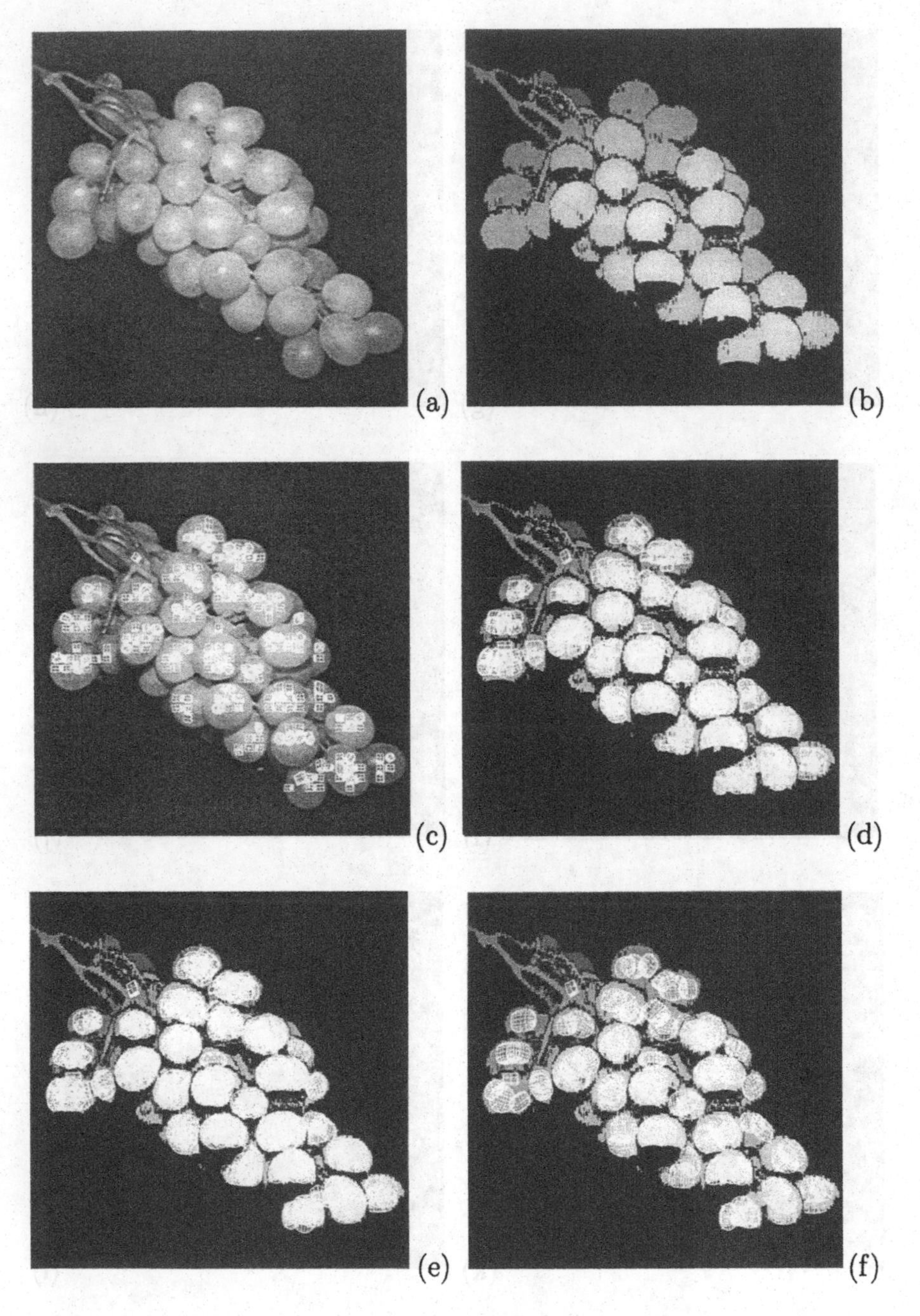

Figure 6.3. Segmentation of a bunch of grapes

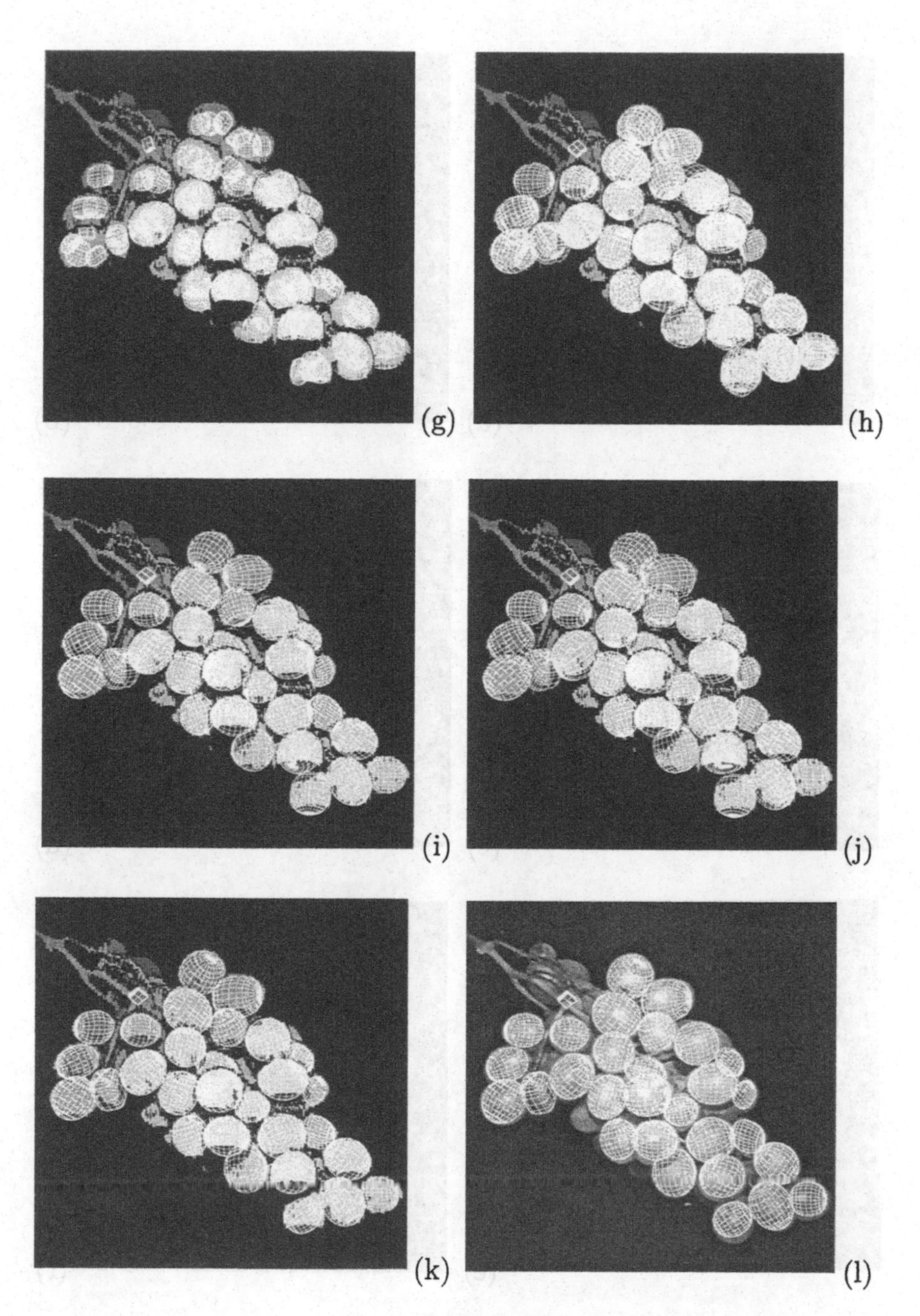

Figure 6.3 (continued).

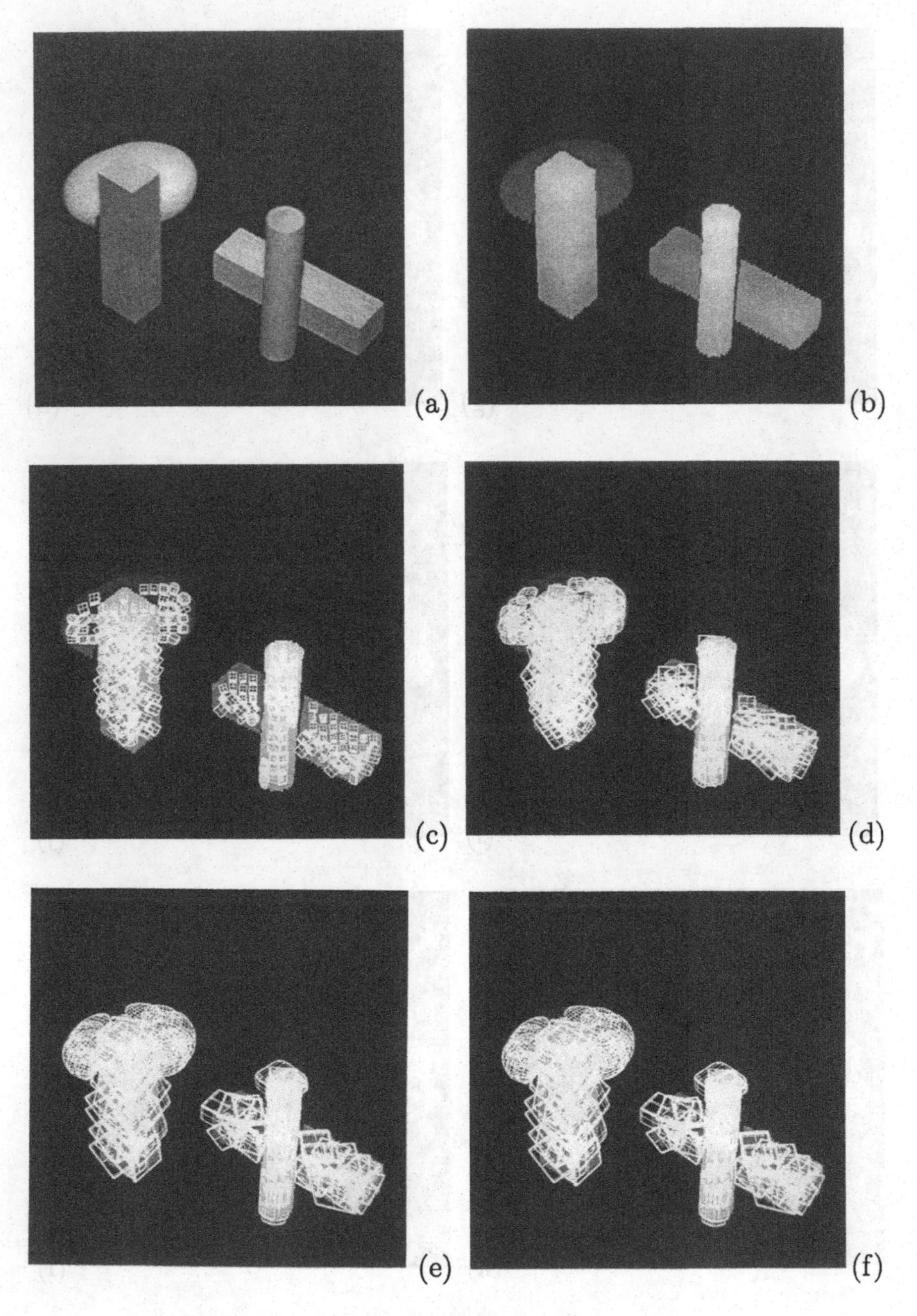

Figure 6.4. Influence of occlusion on segmentation

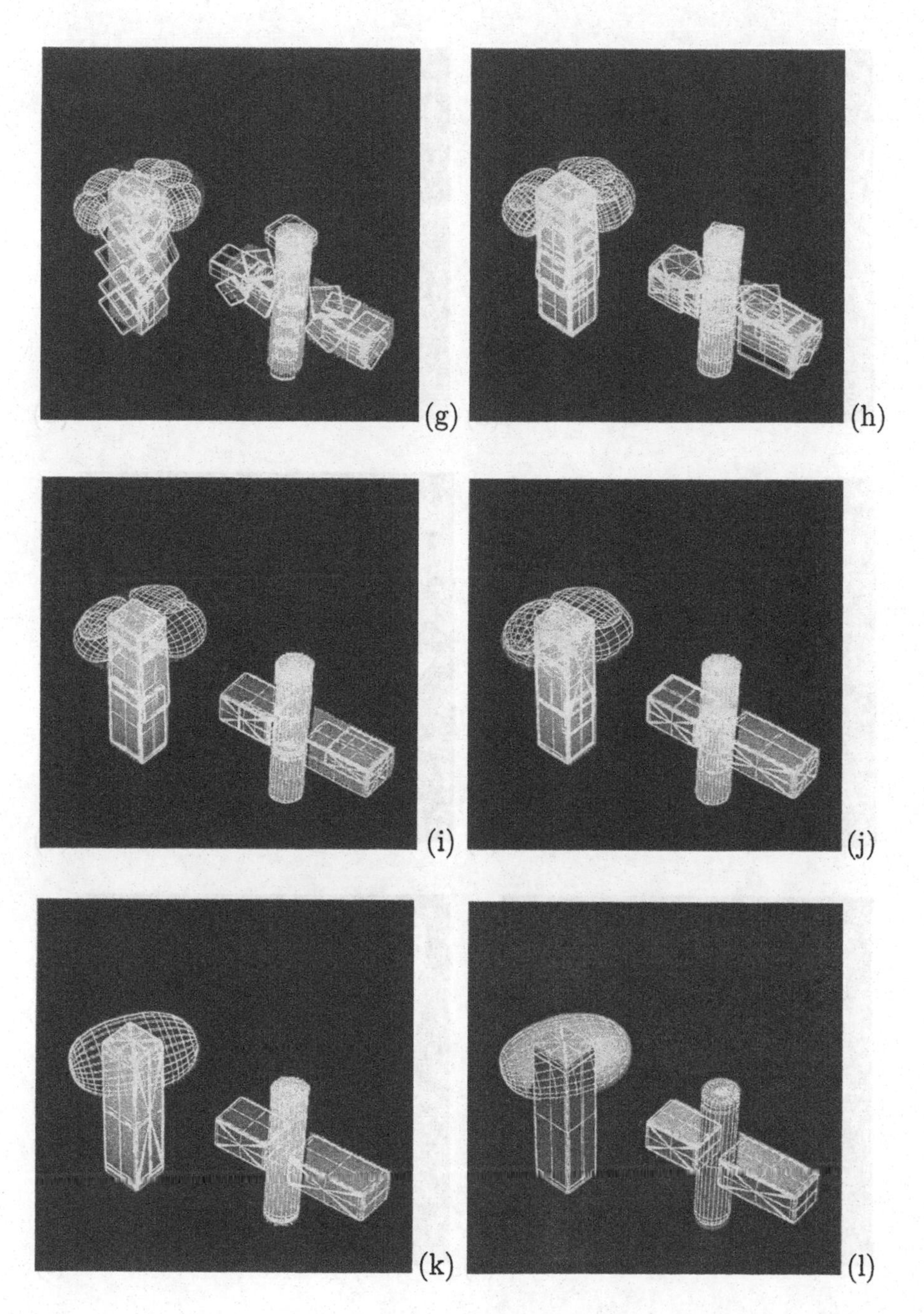

Figure 6.4 (continued).

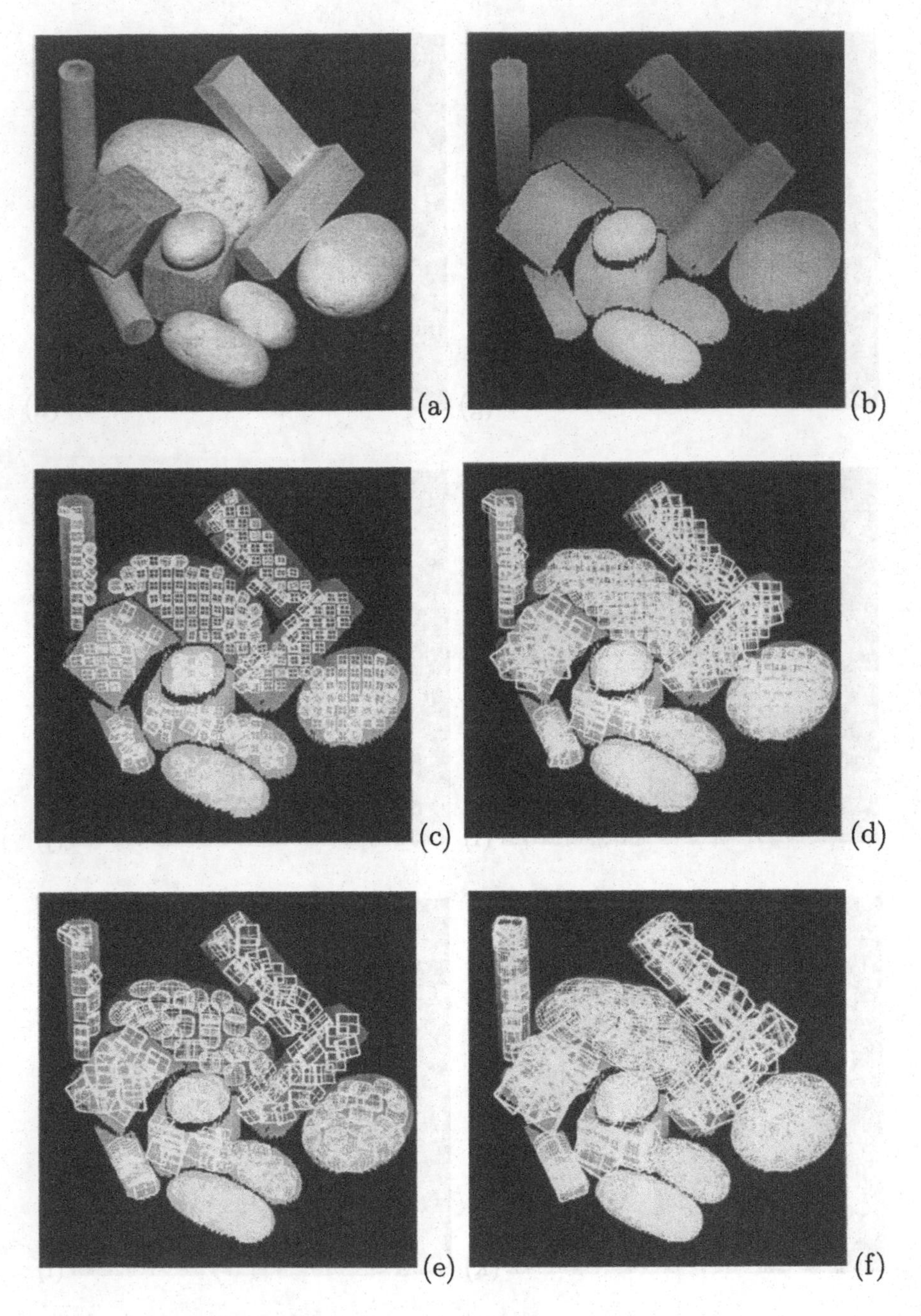

Figure 6.5. Segmentation of a complex scene

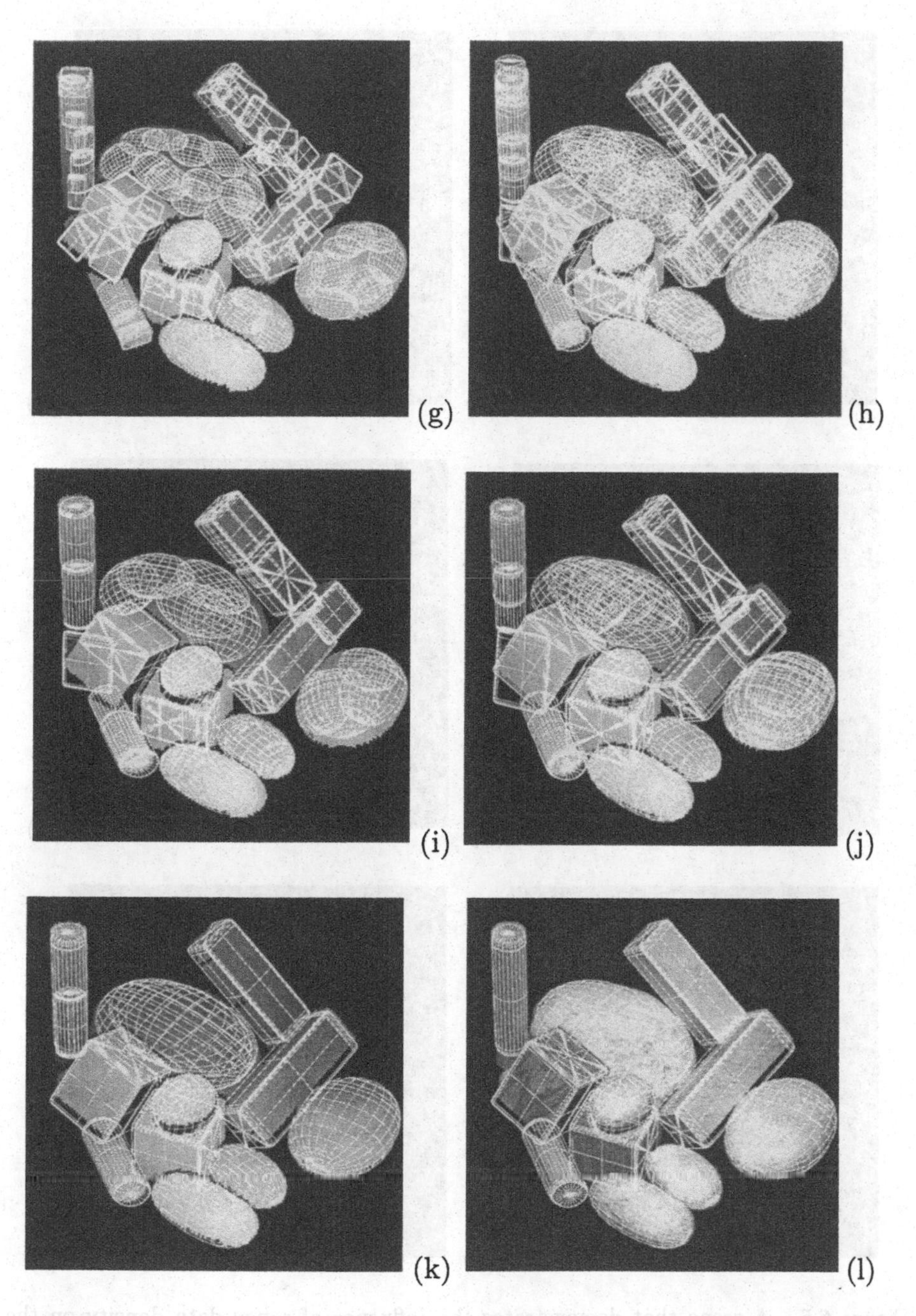

Figure 6.5 (continued).

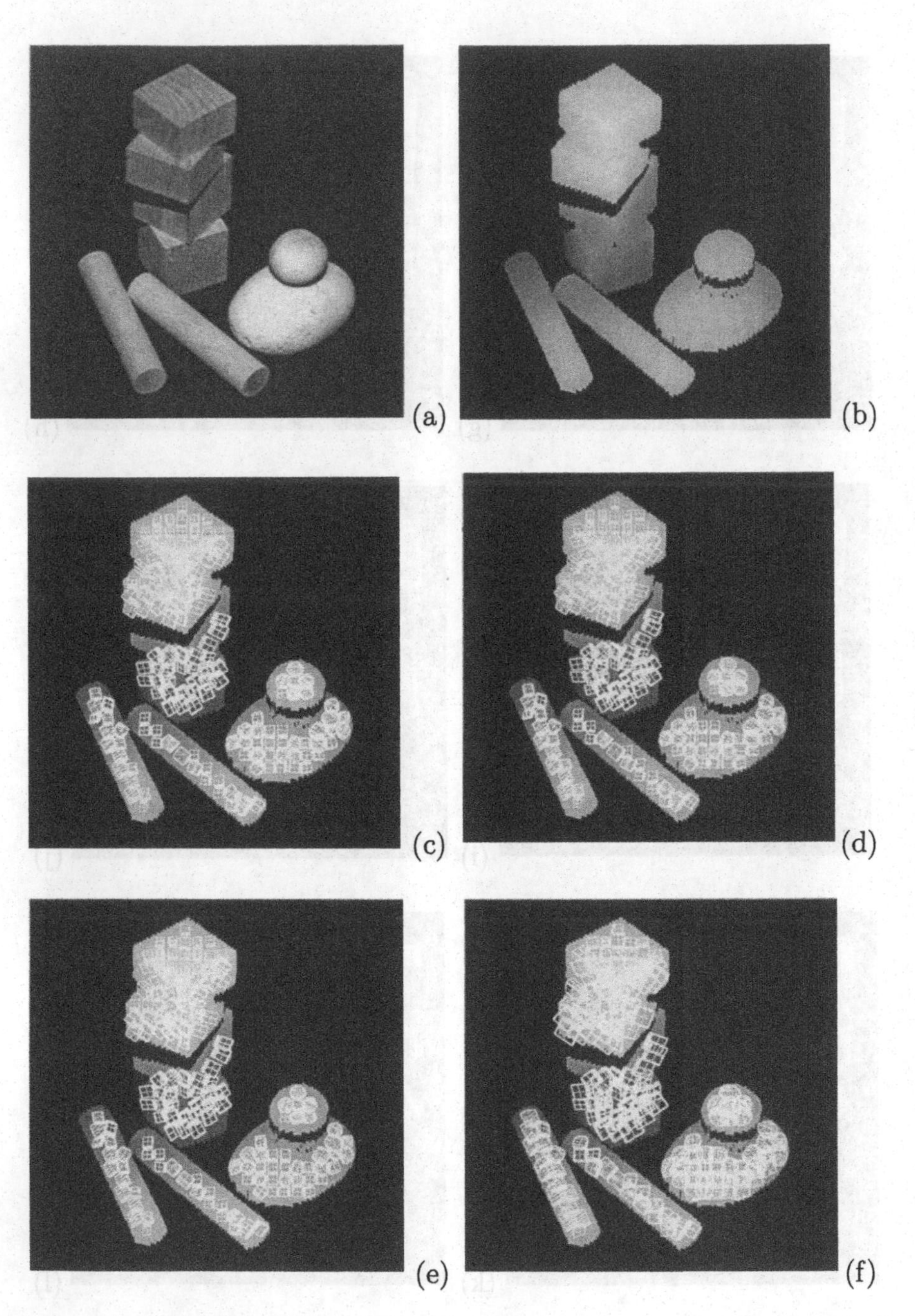

Figure 6.6. A scene that demonstrates the influence of range data density on the recovered models.

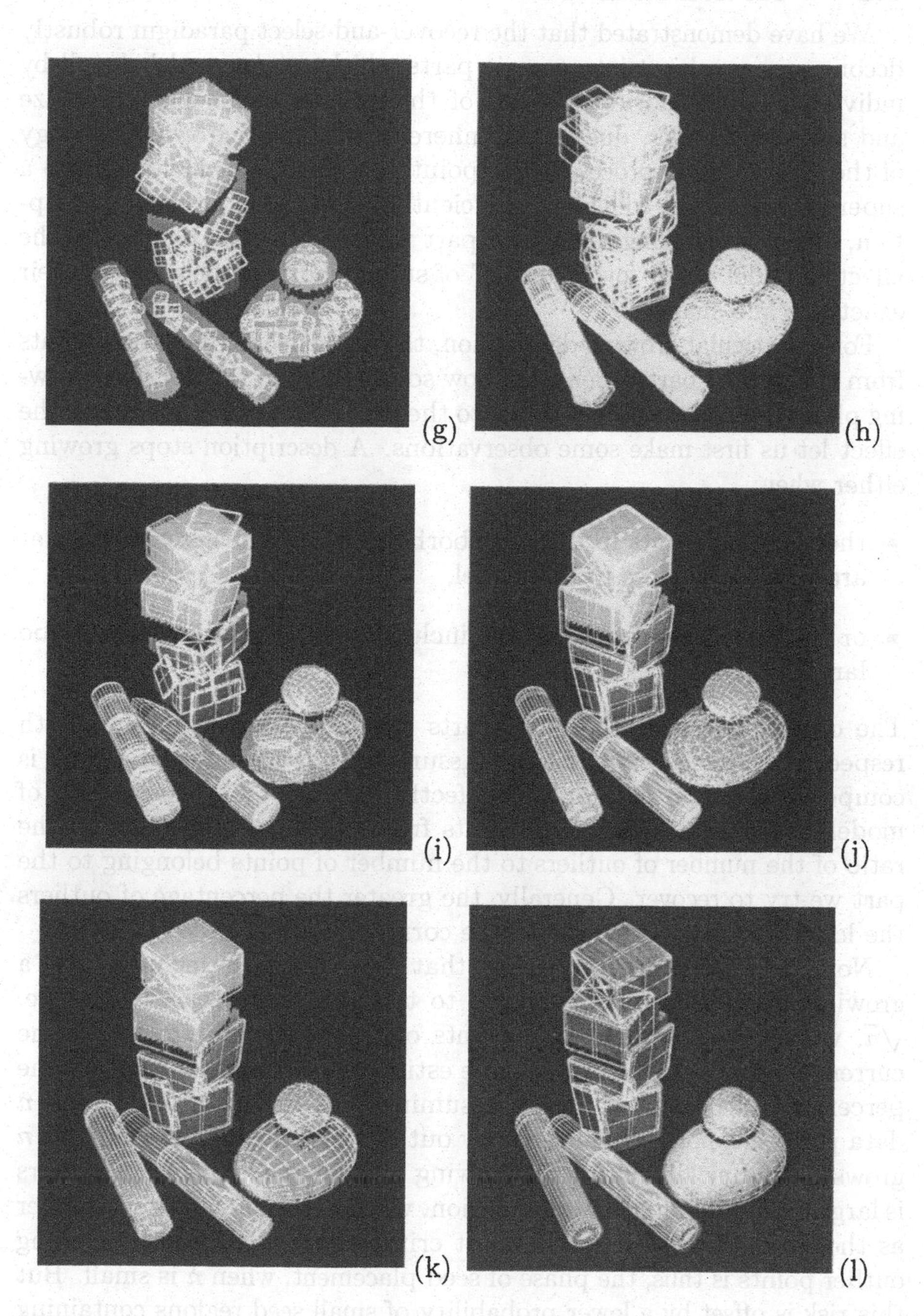

Figure 6.6 (continued).

6.2 ANALYSIS OF RESULTS

We have demonstrated that the recover-and-select paradigm robustly decomposes an object into generic parts which can be modeled well by individual superquadrics in terms of the number of models, their size and shape. However, due to the inherent monotonic growing strategy of the segmentation process, some points are sometimes included into a superquadric model which are sufficiently close to the growing description, but in fact belong to another part which touches or penetrates the affected model. Such "overgrowing" of superquadric models affects their exact size and shape.

For a particular growing description, the final effect of including points from some other part depends on how soon or late in the iterative growing phase the points get included into the description. To understand the effect let us first make some observations. A description stops growing either when:

- there are no points in the neighborhood of current description that are sufficiently close to the model,
- or the error of description that includes the new data points is too large.

The data points from the other parts can be treated as outliers with respect to the recovery process. Assuming that the imaged object is composed of parts that can be perfectly modeled by a chosen type of models, the effect of including points from other parts depends on the ratio of the number of outliers to the number of points belonging to the part we try to recover. Generally, the greater the percentage of outliers the lower the chances to recover the correct model.

Note that the number of points that are potentially included into a growing description is proportional to the perimeter of the region, i.e. $\sqrt{n}$, where n is the number of points or the area of the region in the current description. So a reasonable estimate for the upper limit on the percentage of outliers is $\sqrt{n}/n$, assuming that the description with n data point has not accumulated any outliers so far. Since the number n grows monotonically during the growing process, the influence of outliers is large in the beginning of the iteration, when n is small, and gets smaller as the region gets larger. The most critical part in terms of including outlier points is thus, the phase of seed placement, when n is small. But this risk is offset by a lower probability of small seed regions containing outliers and the large number of redundant seeds. The reasoning about the upper bound of outliers percentage also explains that descriptions eventually stop growing, because there are no points in the neighborhood sufficiently close to the model of the description. When the description approaches its final and correct size, the outliers from the touching or penetrating parts have such a small influence on the recovered model, that the points examined in the next iteration are further and further away from the description's model.

In a real situation the outliers may enter a description continuously. It is easy to find a case where outliers gradually influence the description to such an extent, that even more outliers are included. How can we limit this effect? The effect can be reduced by increasing the range image resolution and consequently lowering the threshold for maximal distance of points that get included into the description during the growth. But the effect can still never be completely eliminated just by increasing the resolution.

To analyze possible alternative solutions, let us make the following two assumptions:

- the scene is composed of parts that can be perfectly modeled by the chosen models (superquadrics),
- the regions where recovered models overlap are small in comparison with the regions contributing to the overlap.

If the overlap regions are sufficiently small, we can reliably recover a model from the data simply by discarding the data points in the overlap regions. Overlapping of regions by itself, strongly suggests that the range points in these regions are not correctly assigned to models. Note that such a conclusion cannot be made simply from the distance of these points from the model, since these distances were limited in advance already during the growing phase. Besides, these data points might have already affected the model that we use for computing the distance.

6.2.1 POST-PROCESSING OF RESULTS

Based on the above analysis, we defined the following qualitative filtering of regions. Each of the superquadric descriptions is processed by the post-processing algorithm in Table 6.2.

As a side effect, this algorithm also guarantees strict partitioning of the input data set, which is required in some applications. Since the assignment of points might break the 8-connectedness of regions enforced by the growing algorithm, we need to perform an addition step of computing the 8-connectedness closure of points from the overlapping regions with respect to the regions, which they were assigned to.

We have analyzed the average errors $(\overline{\xi_i})$ and the errors distribution of final superquadric models for two segmentation examples shown in

Table 6.2. Algorithm for post-processing regions and models

1. From the region we subtract regions of all the other descriptions.
2. A superquadric model is then fitted to this filtered region.
   We expect to get a better model even though the regions are smaller, since they contain less outliers. To prevent running into the same problem, that we are trying to solve, we cannot grow the filtered regions. Instead we assign points from the overlapping subregions to the filtered region with the closest model.
3. We then finally fit superquadric models to these updated regions.

Figs. 6.7 and 6.11. Note that a point inside a superquadric is assigned a negative distance, with the absolute value equal to the distance. A point outside a superquadric is assigned the actual distance value.

To better illustrate the described effects, we have chosen subsampling of 4 on range images of size 256×256 pixels with the values of segmentation parameters shown in Table 6.3. The models recovered by the basic recover-and-select algorithm are compared with the improved models after the post-processing. The post-processing of recovered superquadrics results in a significant improvement of the error distributions for touching or penetrating parts.

Table 6.3. Thresholds for model recovery and constants for model selection

max_point_distance	*max_average_model_error*	K_1	K_2	K_3
6.0	2.0	1.0	0.3	1.0

Superquadric models recovered in Fig. 6.7 and their respective data point sets are shown before post-processing in Fig. 6.8. The results of post-processing are shown in Fig. 6.9. The initial radius of part 1 (a_1 and a_2) is too large because the corresponding superquadric has grown into points which belong to part 2 (Fig. 6.8 (d)). Part 3 is also affected, but to a lesser degree (Fig. 6.8 (f)). After post-processing individual superquadric models, one can clearly observe the improvements in segmentation as a consequence of more correct size and shape estimation of superquadric models (Fig. 6.9 (d)). Fig. 6.10 compares the error distributions of individual superquadrics before and after post-processing. The largest improvement can be observed for part 1.

Fig. 6.11 shows the segmentation results for an L-shaped object. The object is decomposed into two intersecting superquadrics because there is no interaction among superquadrics during the process of growth. Besides preventing bad intermediate model estimates to affect other models, this unconstrained and independent growth of superquadric models offers the additional benefit in terms of the *uniqueness of the decomposition* of the object into parts. If we assume non-protruding blocks for representation, two different decompositions are possible. Fig. 6.12 shows, how as a side effect of the post-processing, a strict partitioning of the range data is achieved according to the recovered models. Fig. 6.13 shows again the comparison of the error distributions before and after post-processing.

An experimental comparison of our superquadric segmentation method to other methods, that attempt to segment range images into superquadric models, is possible in a similar way to the comparison of segmentation algorithms with planar patches (Hoover et al., 1996).

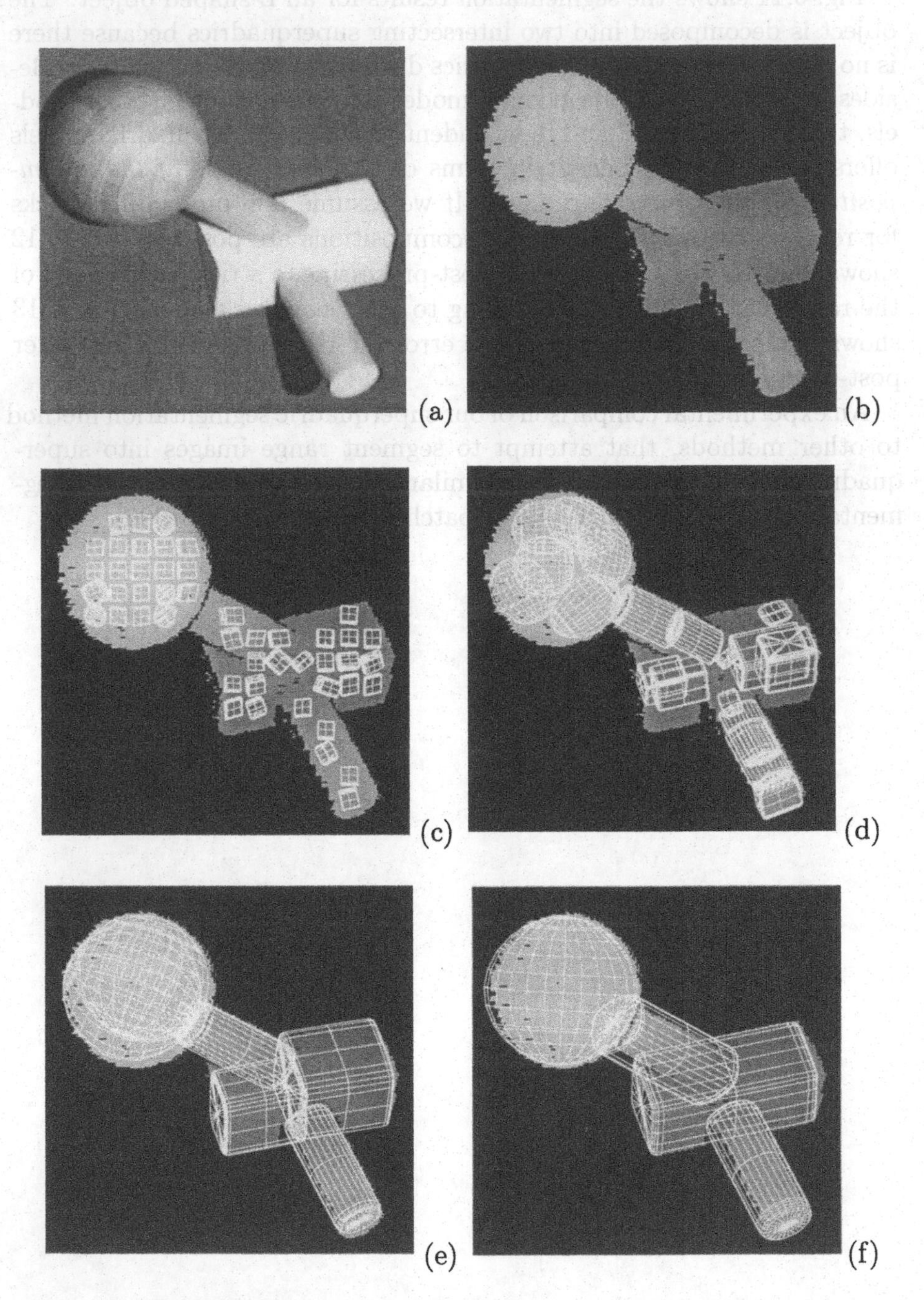

Figure 6.7. Segmentation of an articulated object: (a) intensity image, (b) range image, (c) initial seeds, (d) selection after first growth, (e) selection after second growth, (f) final result of segmentation.

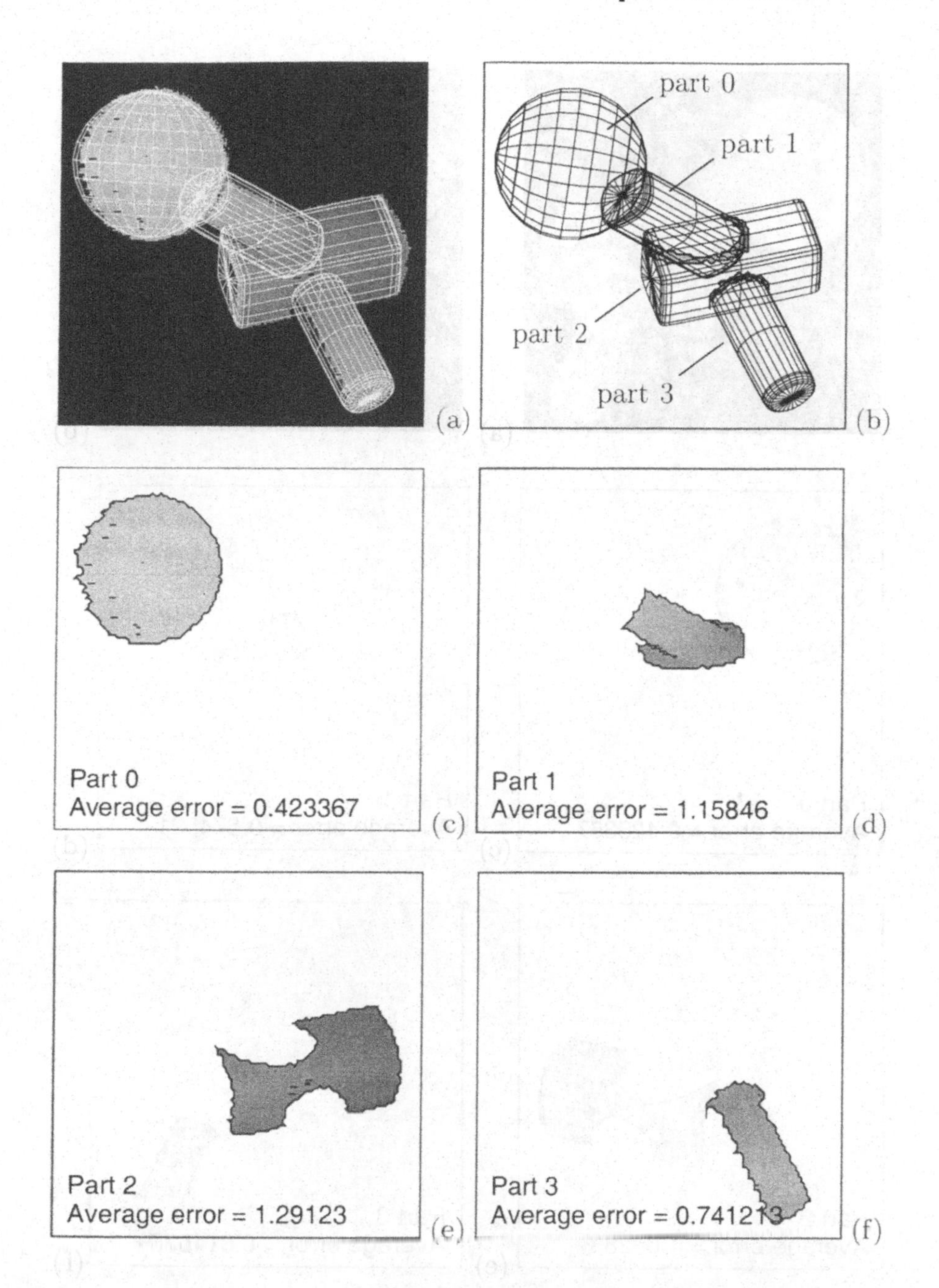

Figure 6.8. Analysis of segmentation results in Fig. 6.7: (a) range image overlaid with the four recovered superquadrics, (b) the two darker regions of points belong to more than one model, (c),(d),(e),(f) regions of range data corresponding to individual superquadric models. The radius (a_1 and a_2) of part 1 is too large because it has grown into points which belong to part 2 (d). Part 3 is also affected but to a lesser degree (f).

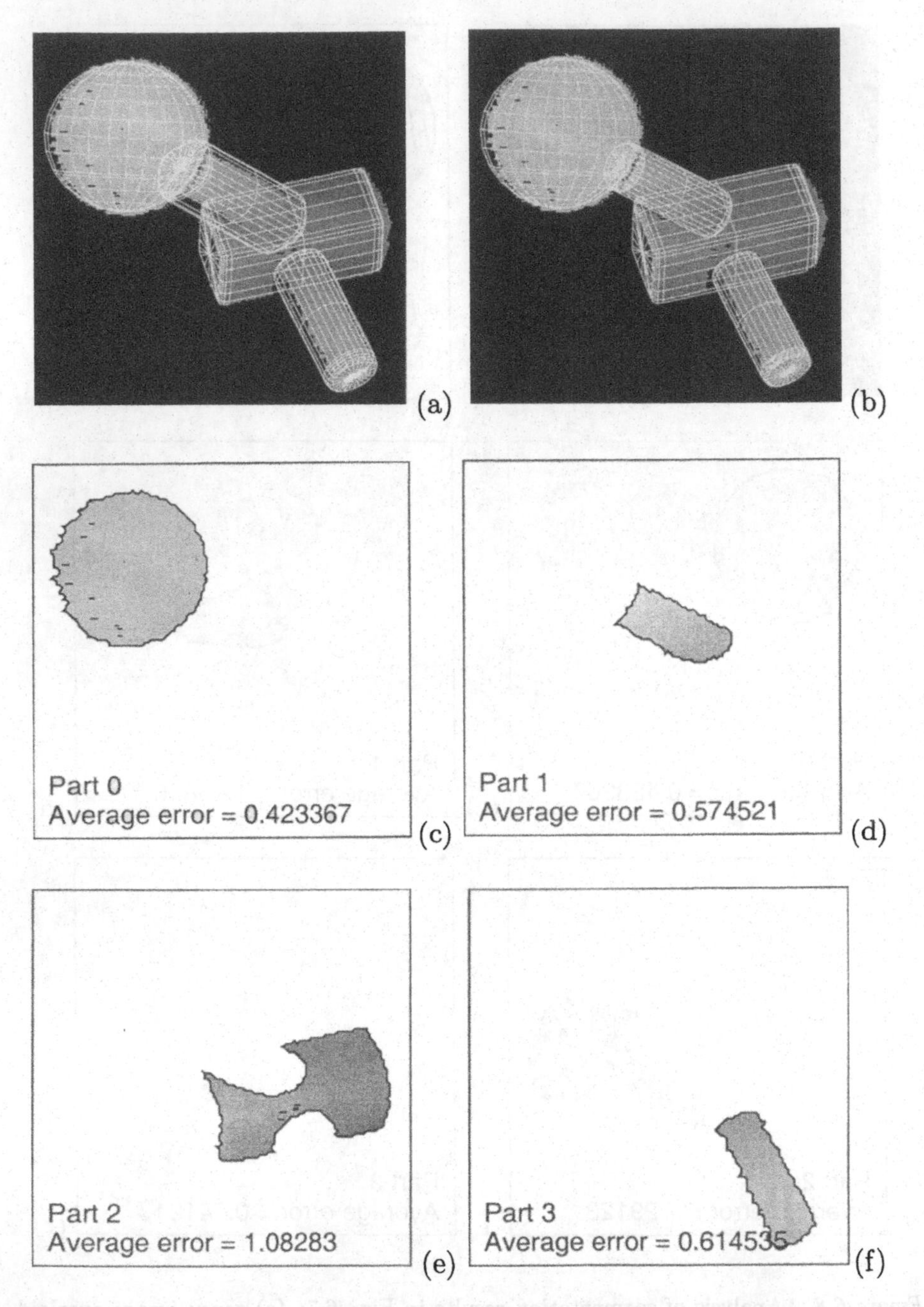

Figure 6.9. Analysis of post-processing the recovered models of the articulated object in Fig. 6.7: (a) superquadric models before post-processing, (b) superquadric models after post-processing, (c),(d),(e),(f) regions of range data corresponding to individual parts after post-processing. Improvement in segmentation and consequently of the correct size of superquadric models for part 1 an part 3 is clearly visible.

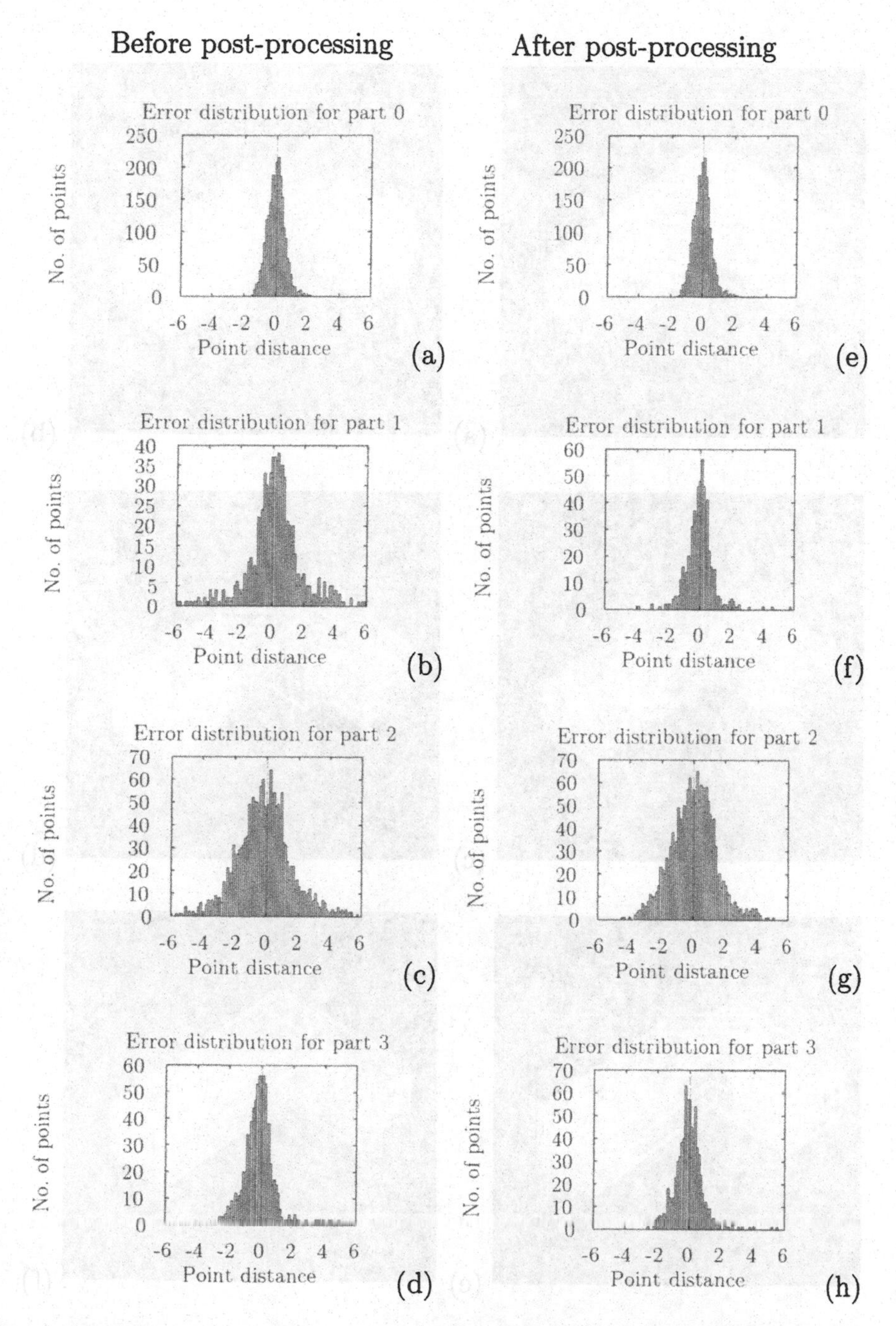

Figure 6.10. Error distributions for superquadric models of parts in Fig. 6.8 before: (a), (b), (c), (d); and after post-processing: (e), (f), (g), (h). Note that a point inside a superellipsoid is assigned a negative distance, with the absolute value equal to the distance. A point outside the superellipsoid is assigned the real distance value.

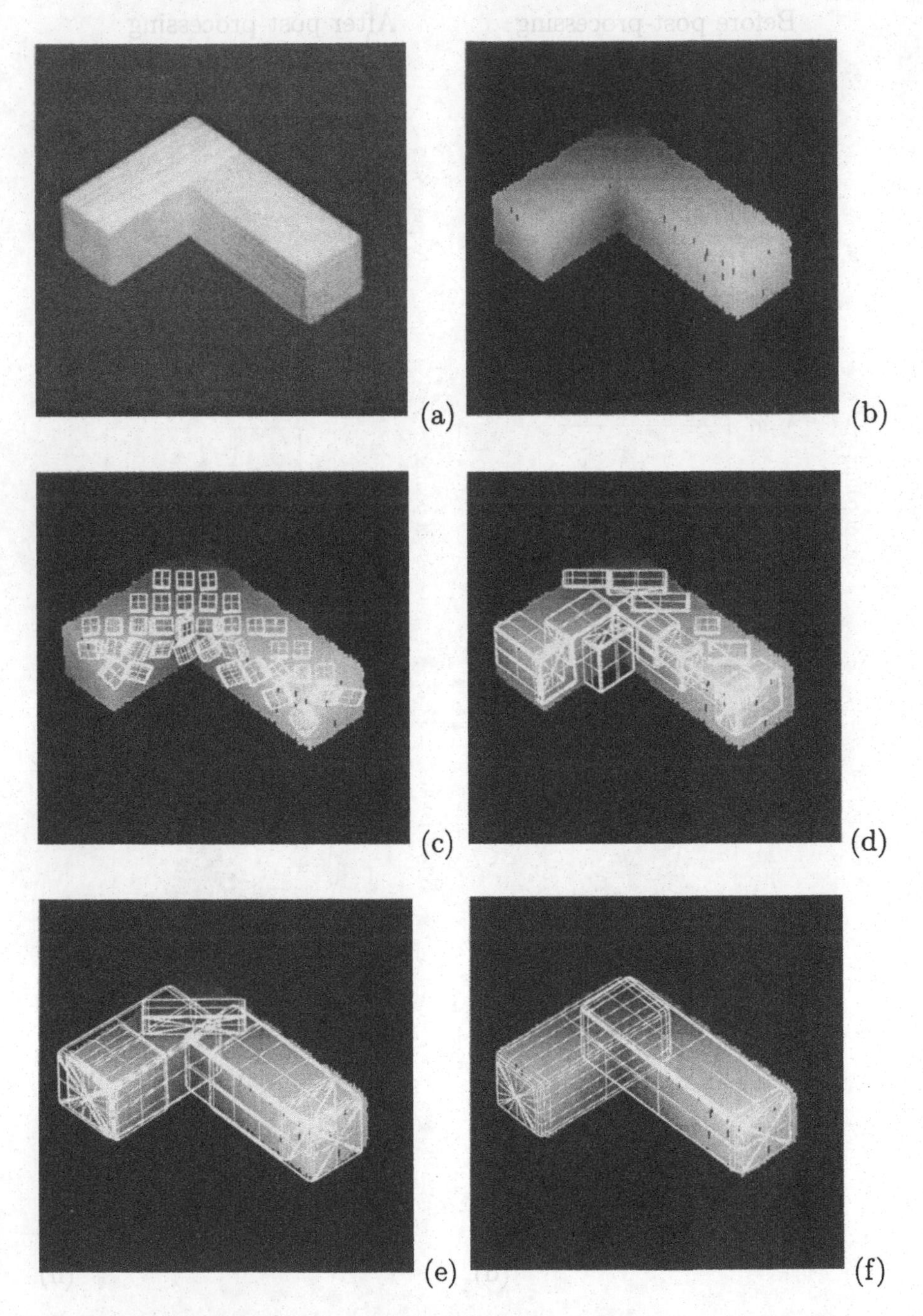

Figure 6.11. Segmentation of an L-shaped object: (a) intensity image, (b) range image, (c) initial seeds, (d) selection after first growth, (e) selection after second growth, (f) final result of segmentation.

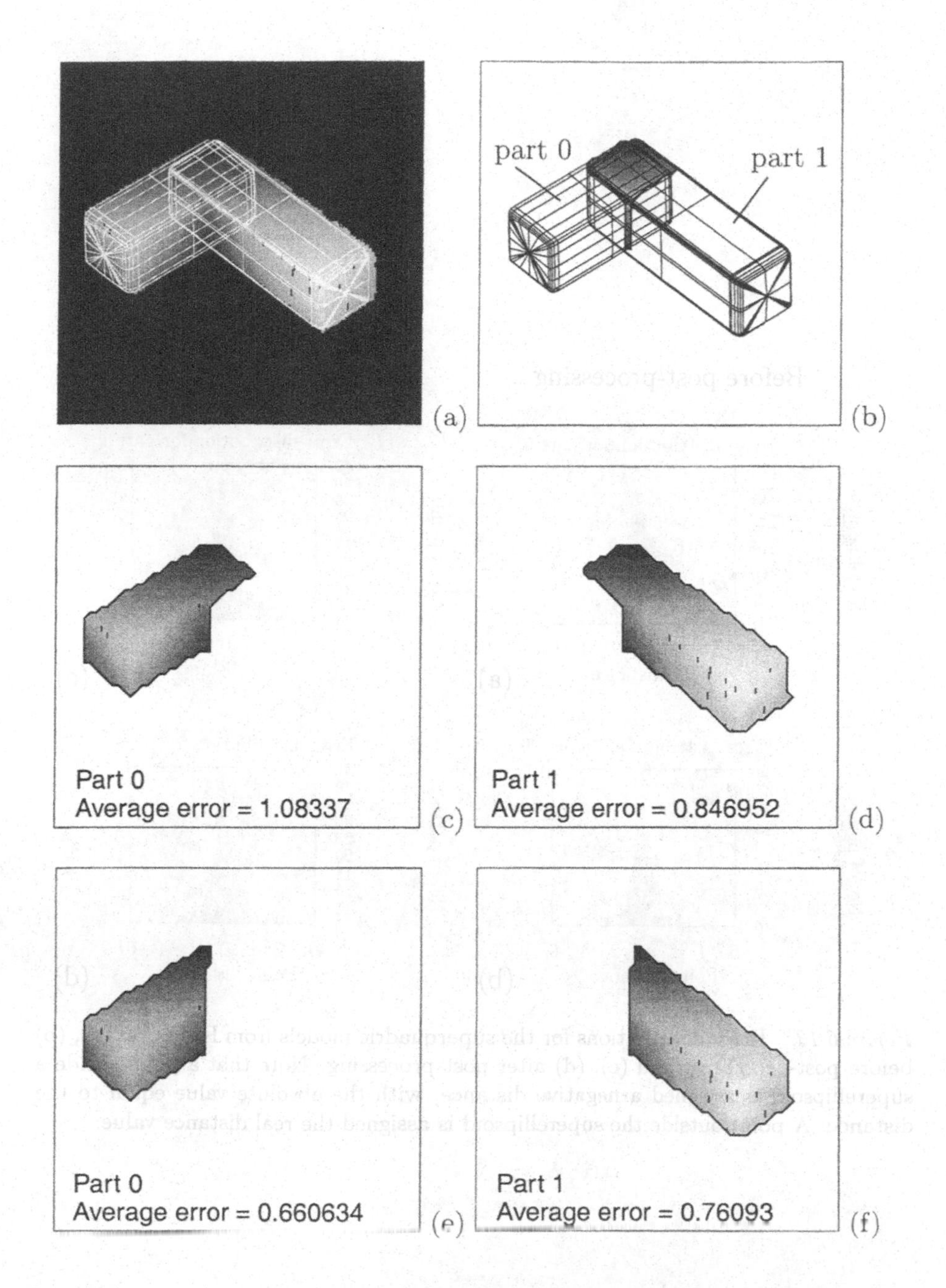

Figure 6.12. A side effect of the post-processing step is also a strict partitioning of the range data. The effect is shown on the results of segmentation in Fig. 6.11: (a) shows the superquadric models after post-processing; (b) the darker region indicates points that belong to both models; (c) and (d) show the partitioning of the range data after the basic segmentation algorithm; (e) and (f) show the partitioning after post-processing.

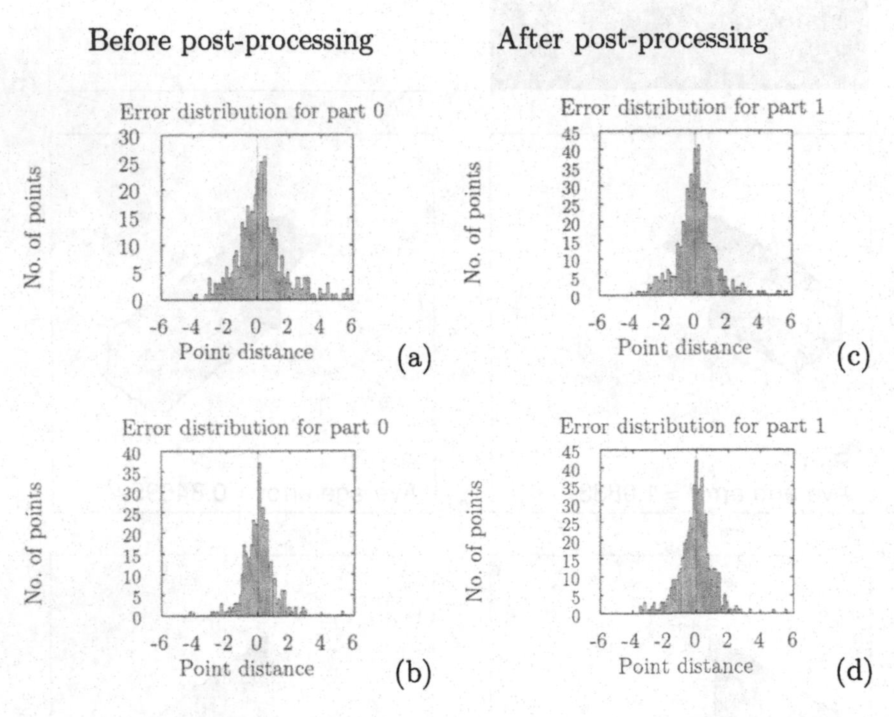

Figure 6.13. **Error distributions for the superquadric models from Fig. 6.12: (a), (b) before post-processing and (c), (d) after post-processing. Note that a point inside a superellipsoid is assigned a negative distance, with the absolute value equal to the distance. A point outside the superellipsoid is assigned the real distance value.**

6.2.2 STABILITY OF SEGMENTATION WITH RESPECT TO CHANGING VIEWPOINT

In this section we demonstrate the stability of segmentation with respect to the number of recovered superquadric models and their shape from different view points. We generated a set of synthetic range images of the same CAD model seen from different viewpoints (Fig. 6.14). We segmented each range image with our segmentation algorithm and compared the results.

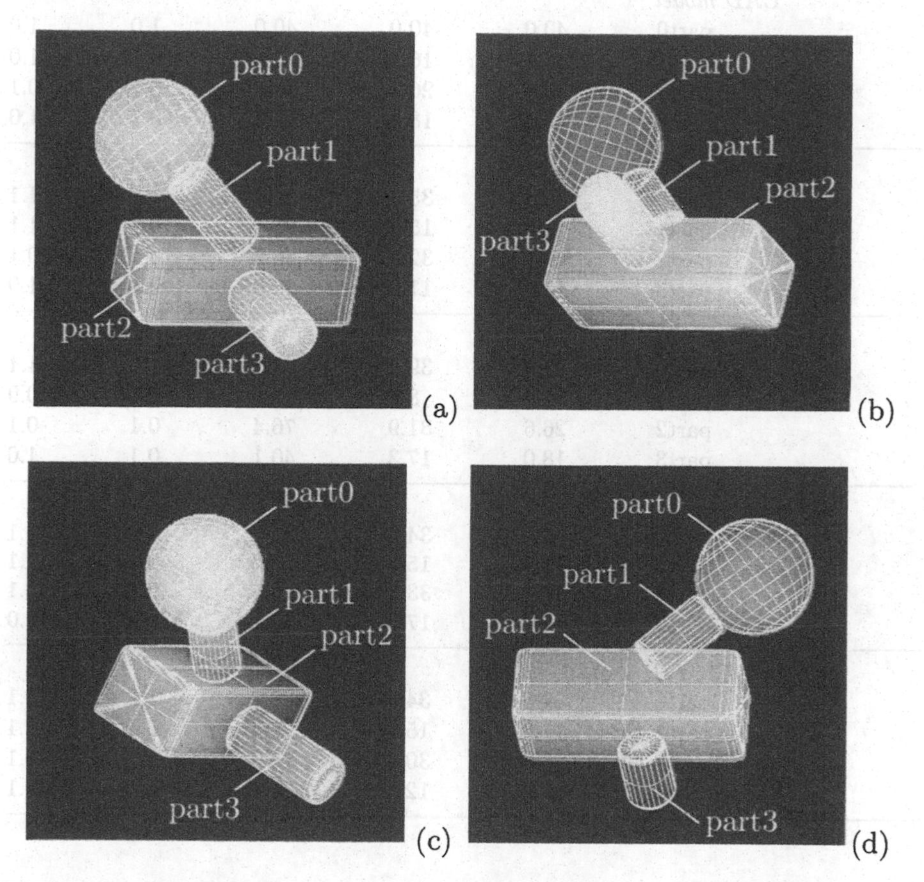

Figure 6.14. Segmentation of objects that can be perfectly represented with superquadrics is stable with respect to the changing viewpoint even in the presence of moderate self-occlusion.

The comparison of segmentation results in Fig. 6.14 shows that our method correctly identifies individual parts even in the case of significant occlusion. Part 1 in Fig. 6.14 (b) is, by a large degree, occluded by parts 2 and 3. Nevertheless, the resulting model for part 1 is still of cylindric shape, but somewhat shorter and thinner. Table 6.4 gives a comparison of the recovered superquadric parameters $(a_1, a_2, a_3, \varepsilon_1, \varepsilon_2)$

from each range image in Fig. 6.14 with the parameters of the original CAD model used for generation of range images.

Table 6.4. Comparison of superquadric parameters recovered from range images of the same object from different viewpoints (Fig. 6.14) with the parameters of the original CAD model.

Superquadric parameters	a_1	a_2	a_3	ε_1	ε_2
CAD model					
part0	40.0	40.0	40.0	1.0	1.0
part1	16.7	16.7	50.0	0.1	1.0
part2	33.3	26.7	76.7	0.1	0.1
part3	18.0	18.0	41.7	0.1	1.0
Figure 6.14 (a)					
part0	39.4	35.0	39.1	1.0	1.1
part1	12.2	15.8	35.1	0.1	1.1
part2	26.1	32.0	76.2	0.1	0.1
part3	17.9	17.8	39.9	0.1	1.0
Figure 6.14 (b)					
part0	33.8	39.2	38.9	1.0	1.1
part1	2.2	8.8	23.8	0.1	0.9
part2	26.6	31.9	76.4	0.1	0.1
part3	18.0	17.3	40.1	0.1	1.0
Figure 6.14 (c)					
part0	39.2	34.8	39.0	1.0	1.1
part1	11.9	15.8	27.2	0.1	1.1
part2	26.4	33.5	75.6	0.1	0.1
part3	16.7	17.8	39.1	0.1	1.0
Figure 6.14 (d)					
part0	39.8	34.0	38.9	1.0	1.1
part1	11.2	15.5	27.4	0.1	1.1
part2	26.6	30.2	76.5	0.1	0.1
part3	16.7	12.5	34.6	0.1	1.1

6.2.3 SENSITIVITY OF SEGMENTATION TO OUTLIERS

Real range images usually contain large measurement errors which are known as outliers. Outliers appear especially often at depth discontinuities. Such outliers are difficult to observe in grey level coded range images. We can see them better if we look at the 3D range points from a different viewpoint. Fig. 6.15 (a) shows the range points from the range image shown in Fig. 6.7 after a slight rotation. Outliers can now be clearly seen at depth discontinuities.

Fig. 6.15 (b), on the other hand, shows from the same viewpoint only those data points which were included into the recovered superquadric models (Fig. 6.7). The outlier range points are now missing which shows that the growing phase of the recover-and-select paradigm can successfully eliminate outlier points. This filtering effect can be explained by the constant monitoring of the distance of candidate data points from the current model during the segmentation.

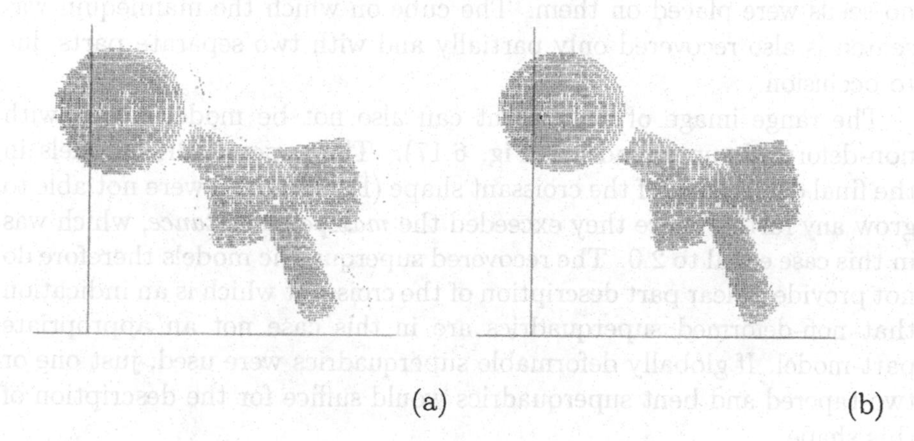

Figure 6.15. Segmentation with the recover-and-select paradigm implicitly filters outliers. From a different viewpoint are shown: (a) range points of the original range image where outliers can be observed, (b) range points which were included into the recovered superquadric models.

6.2.4 SEGMENTATION OF OBJECTS THAT CANNOT BE MODELED WELL BY SUPERQUADRICS

During the course of experimentation, we observed that our segmentation method degrades gracefully if the assumption which is made by the choice of the shape primitives is not met. We can clearly expect good results if the part shapes in the processed range image can be modeled well with superquadrics. This we amply demonstrated with experiments shown in Figs. 6.2–6.6.

A human figure can be still relatively well partitioned by our method although the limbs of the wooden mannequin in Fig. 6.16 cannot be ideally modeled with superquadrics shapes. All parts of the body are reasonably well modeled. The head and the torso are partitioned very clearly. The limbs, unfortunately, are modeled by several intersecting superquadric models which obscures, somewhat, the natural part structure. The feet of the wooden mannequin are also not modeled because no seeds were placed on them. The cube on which the mannequin was seated is also recovered only partially and with two separate parts due to occlusion.

The range image of a croissant can also not be modeled well with non-deformed superquadrics (Fig. 6.17). The superquadric models in the final description of the croissant shape (Fig. 6.17 (i)) were not able to grow any further since they exceeded the *max_point_distance*, which was in this case equal to 2.0. The recovered superquadric models therefore do not provide a clear part description of the croissant which is an indication that non-deformed superquadrics are in this case not an appropriate part-model. If globally deformable superquadrics were used, just one or two tapered and bent superquadrics would suffice for the description of this shape.

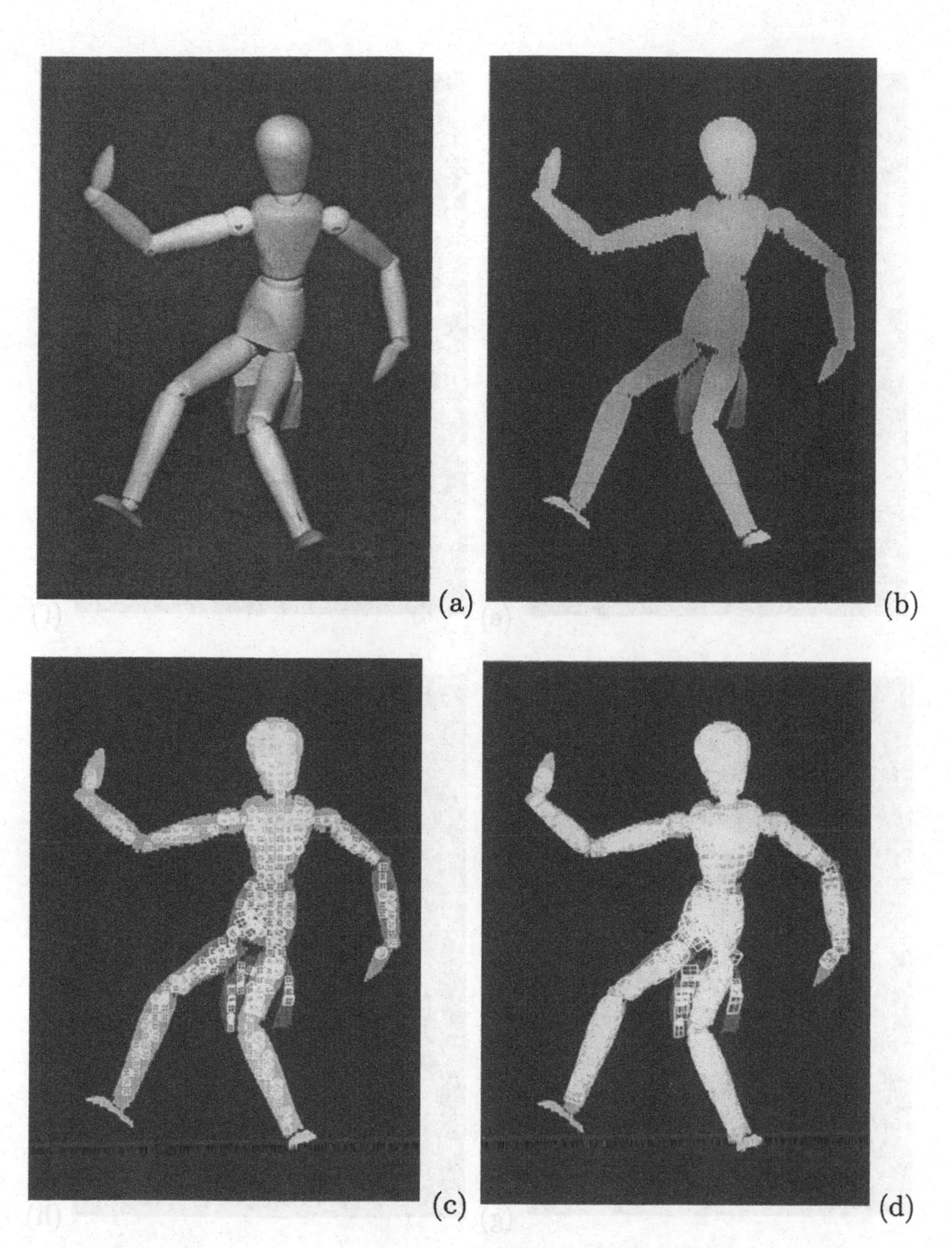

Figure 6.16. Segmentation of a wooden mannequin

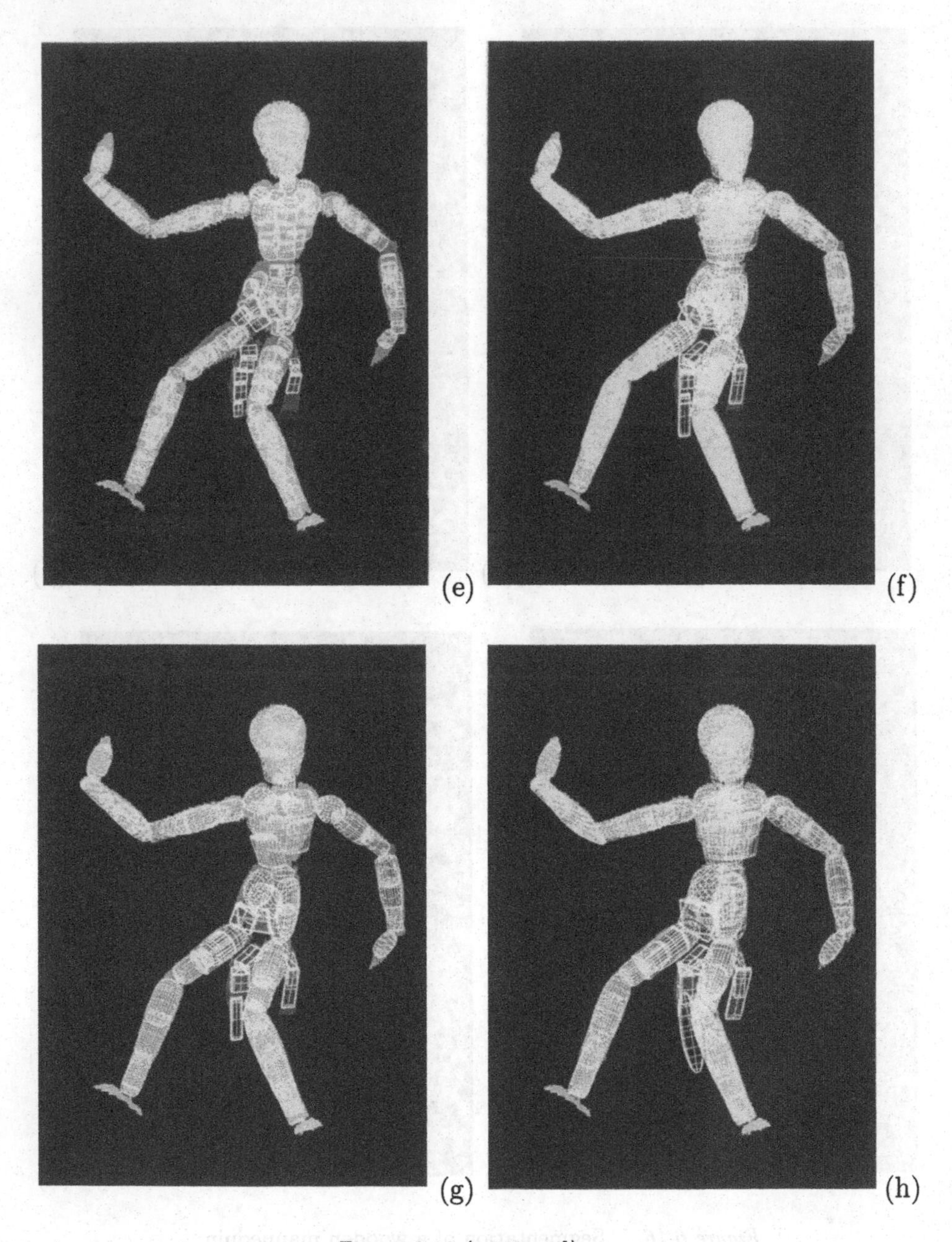

Figure 6.16 (continued).

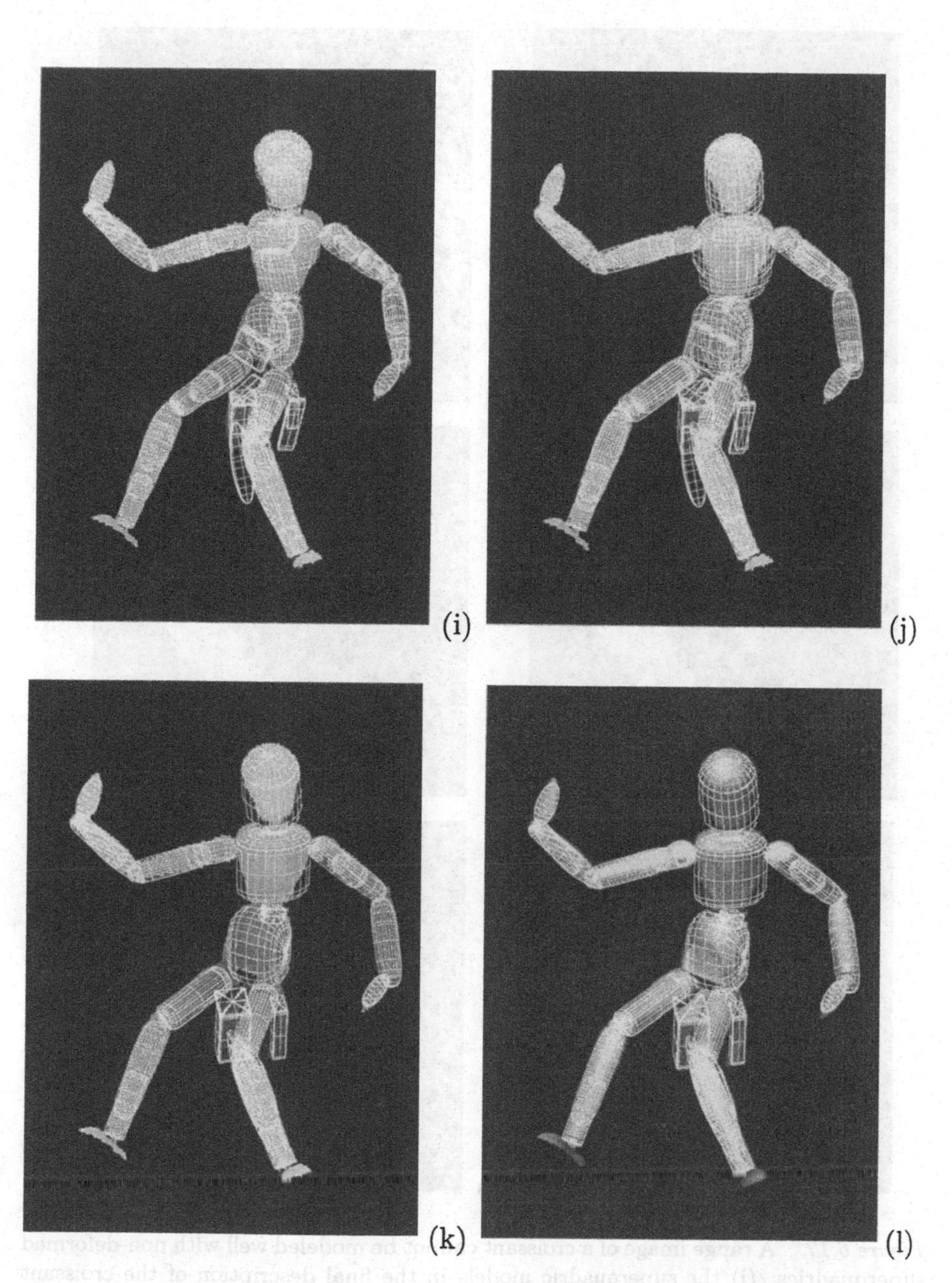

Figure 6.16 (continued).

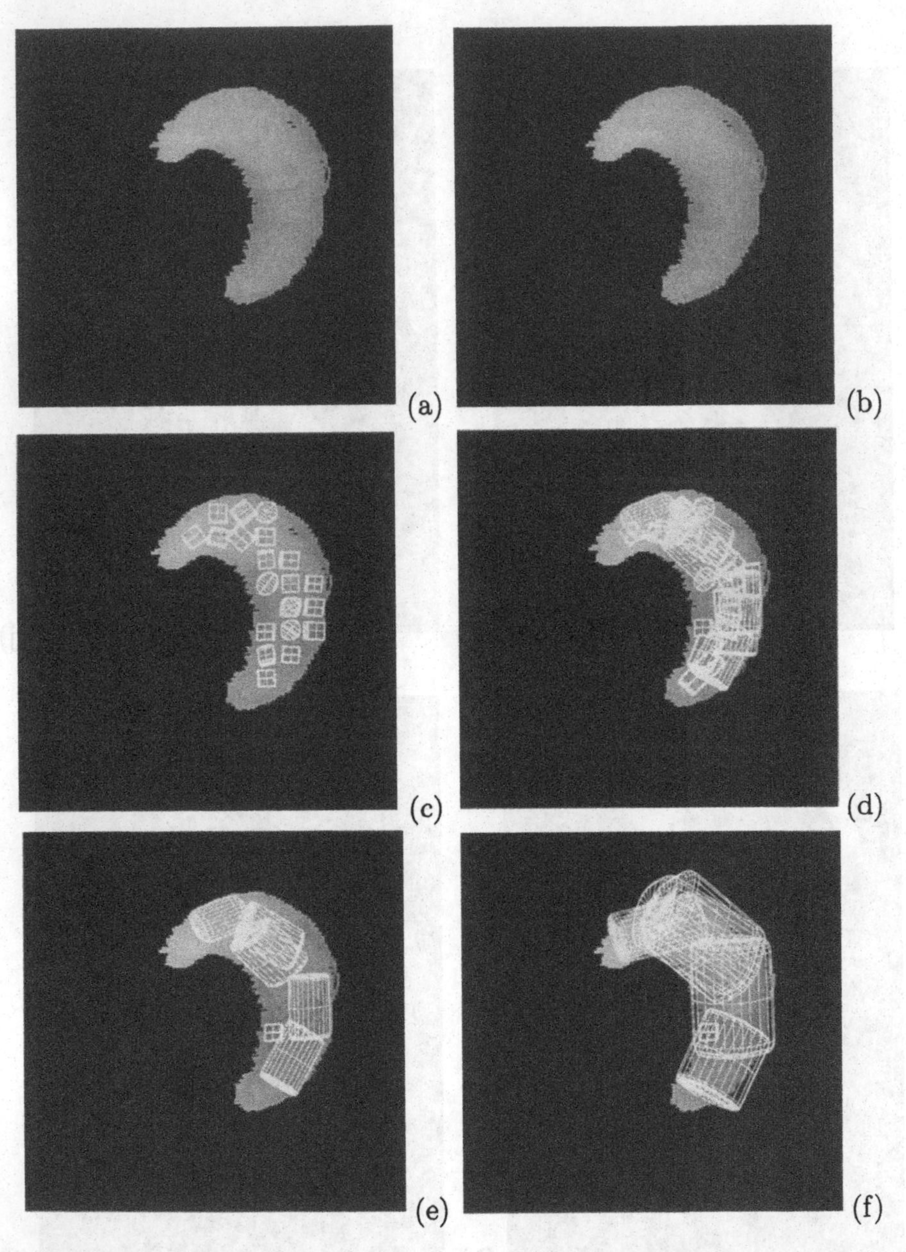

Figure 6.17. A range image of a croissant cannot be modeled well with non-deformed superquadrics: (i) the superquadric models in the final description of the croissant shape were not able to grow any further since they exceeded the maximal allowed average model error.

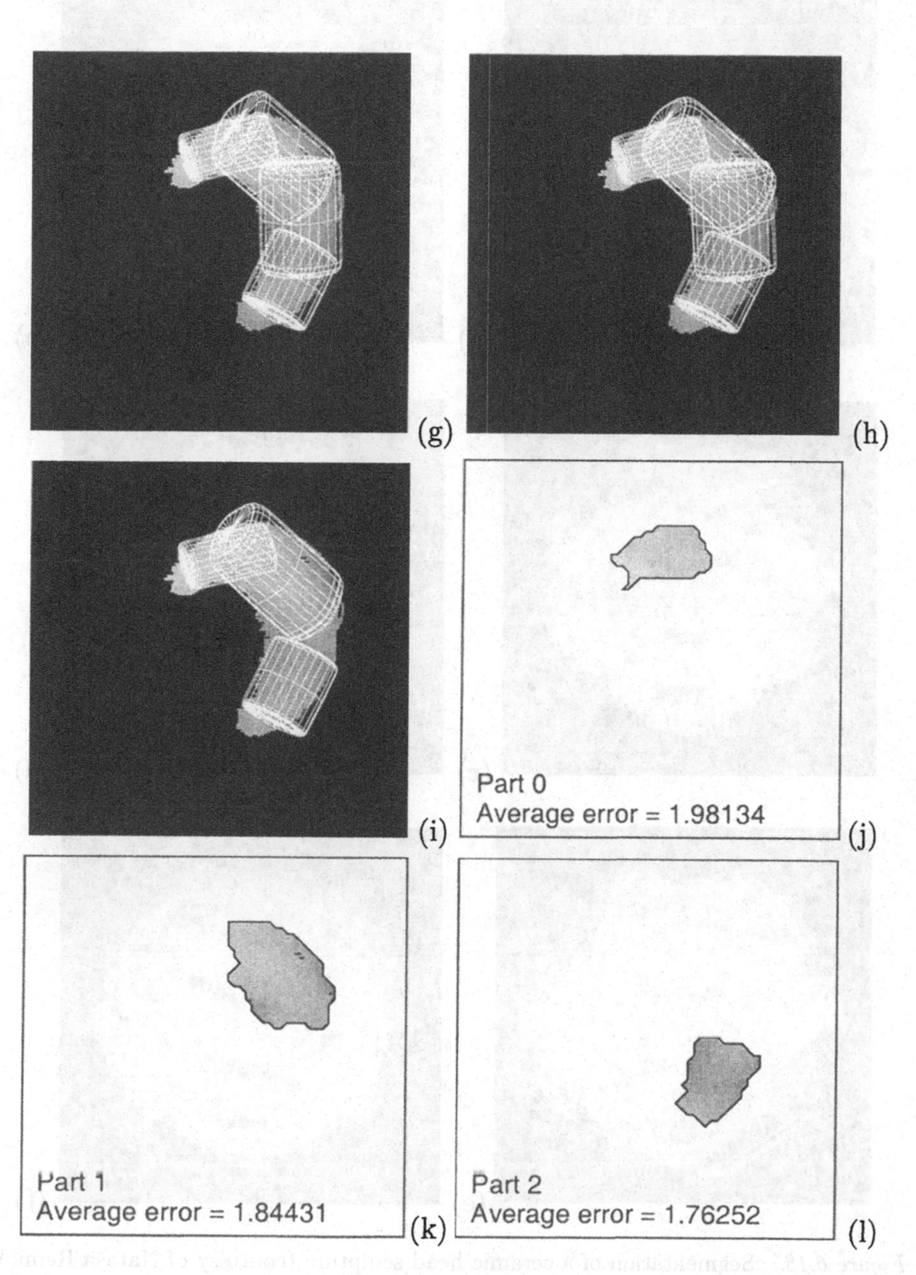

Figure 6.17 (continued).

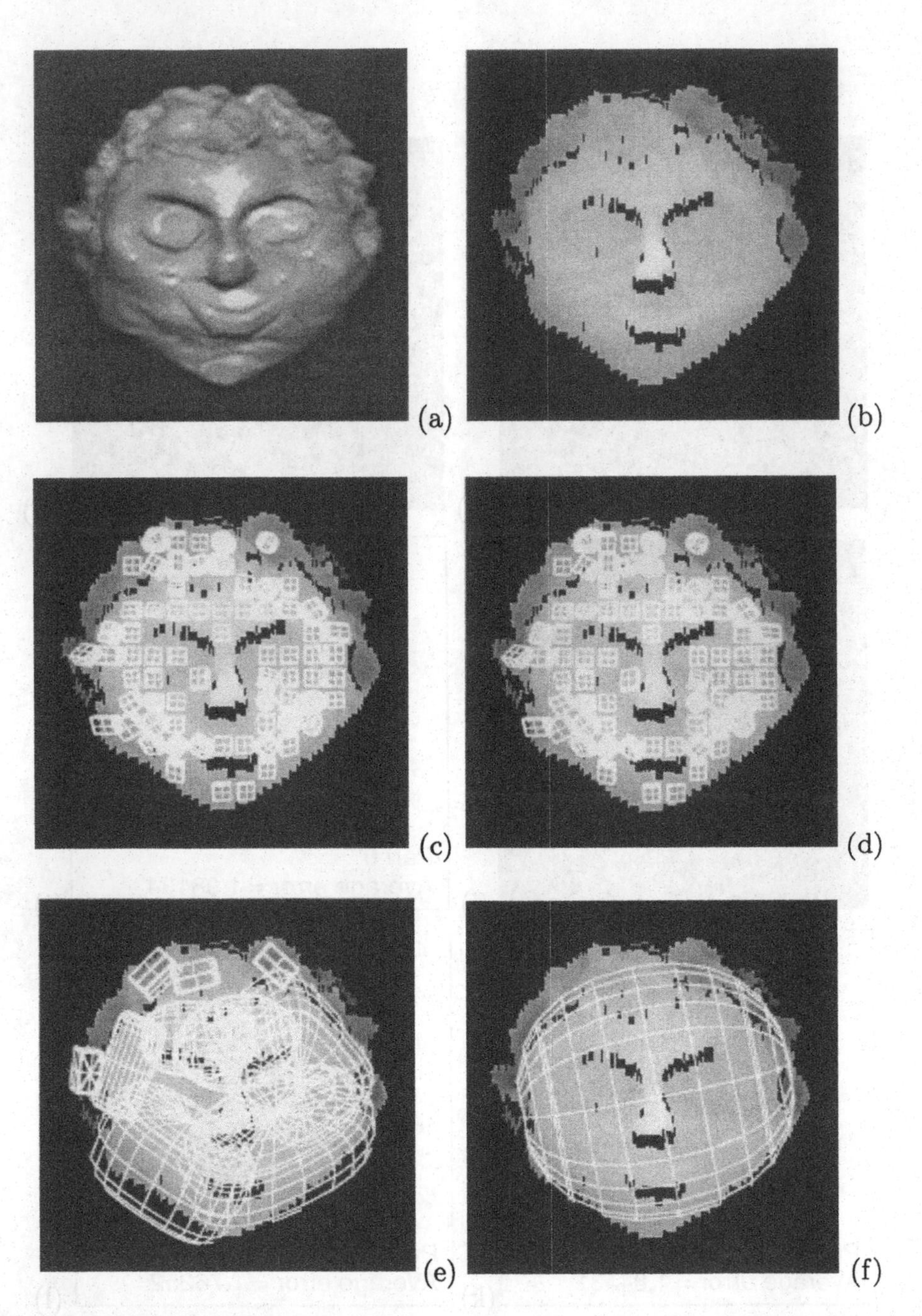

Figure 6.18. Segmentation of a ceramic head sculpture (courtesy of Nataša Remic): (a) intensity image, (b) range image, (c) and (d) initial seeds, (e) recovered superquadric models when *max_point_distance* = 2.1 and *max_average_model_error* = 1.0, (f) recovered superquadric model when *max_point_distance* = 5.0 and *max_average_model_error* = 4.0.

The clay head sculpture in Fig. 6.18 has a very rough surface. If we use the same settings in our segmentation method as in the previous segmentation examples, we get the result shown in Fig. 6.18 (e). If we, however, allow a larger error-of-fit for the superquadrics models, we get the result shown in Fig. 6.18 (f). Since the growth of superquadrics is now less restrictive, only one superquadric models the entire head. This simple example illustrates that there exists a whole hierarchy of different levels of part descriptions. A hierarchy of human body parts, for example, is illustrated in Fig. 1.2. For intelligent interaction with the real world, we need to recover the part structure not only at a single description level, but the whole hierarchy of such descriptions (Solina and Leonardis, 1995). One would like to include in our segmentation method, a generalization or abstraction mechanism which would enable a selection of a particular shape abstraction level. This would make it possible that only parts of a particular scale or a specific coarseness level would be recovered.

6.2.5 SEGMENTATION WITH A COMBINATION OF DIFFERENT MODELS

For representation of visual information it is in general not sufficient to have just a single type of model. In addition to the type of visual information, the selection of an appropriate modeling type depends also on the goal of the vision system which influences the scale of the model, and the processing constraints. To address the selection of appropriate models, we argued for a selective reconstruction of specific shape models tailored to particular functions (Solina and Leonardis, 1998). The image information itself should determine the applicability of the models. Beside specific task-dependent constraints, we used the Minimum Description Length (MDL) principle to select models of appropriate scale among possible different types of models. The MDL principle can automatically determine the scale on which visual phenomena should be observed. Besides the shortest encoding of model parameters, the amount of allowable deviation of visual data from the models is important, which results in interpretations which encompass models of different scales or different types. A particular decision as to which representation to use, rests on how large the deviations from the model are that we can tolerate.

In particular, a scene which can not be modeled accurately enough only with superquadrics can be modeled in a most compact fashion with a combination of superquadrics and surface models (Leonardis et al., 1996). A description consisting only of surface models would be longer while the description consisting only of superquadrics would not be accu-

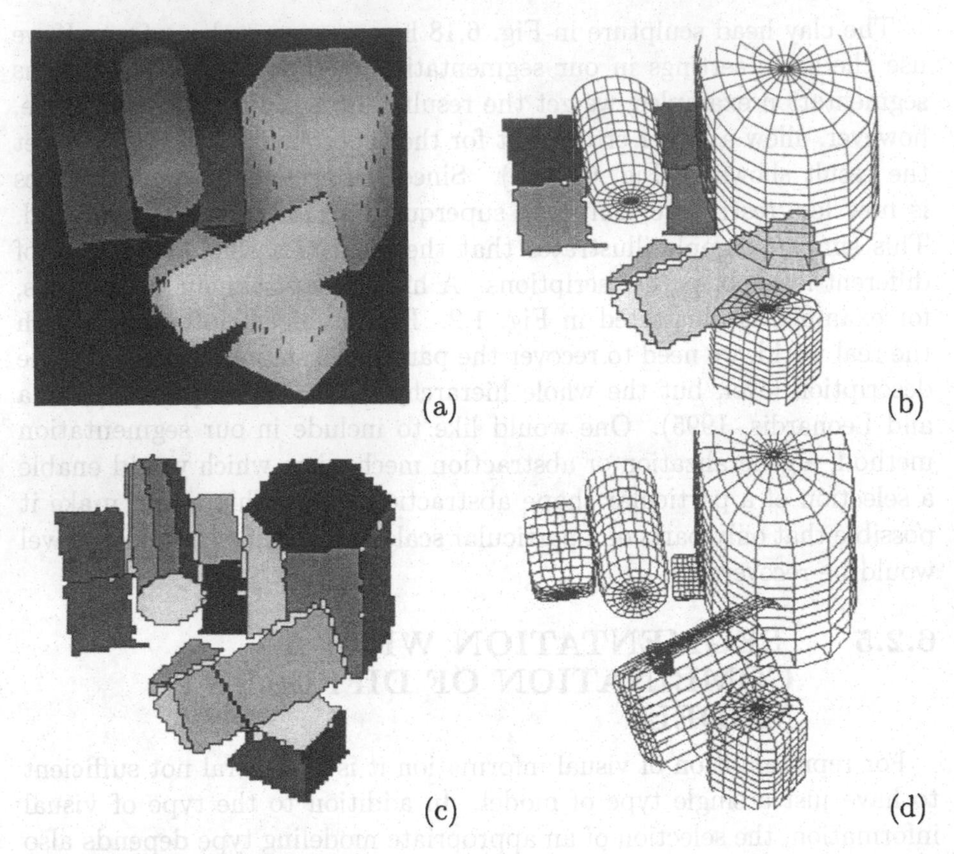

Figure 6.19. Segmentation with volumetric and surface models: (a) original range image, (b) a combined superquadric and planar surface segmentation, (c) independently recovered surface models, (d) independently recovered superquadric models.

rate enough. Fig. 6.19 shows such a combined representation of a range scene. Fig. 6.20 shows the intermediate stages of the recover-and-select segmentation which produced the final results, combining both planar surface models and superquadric models shown in Fig. 6.19.

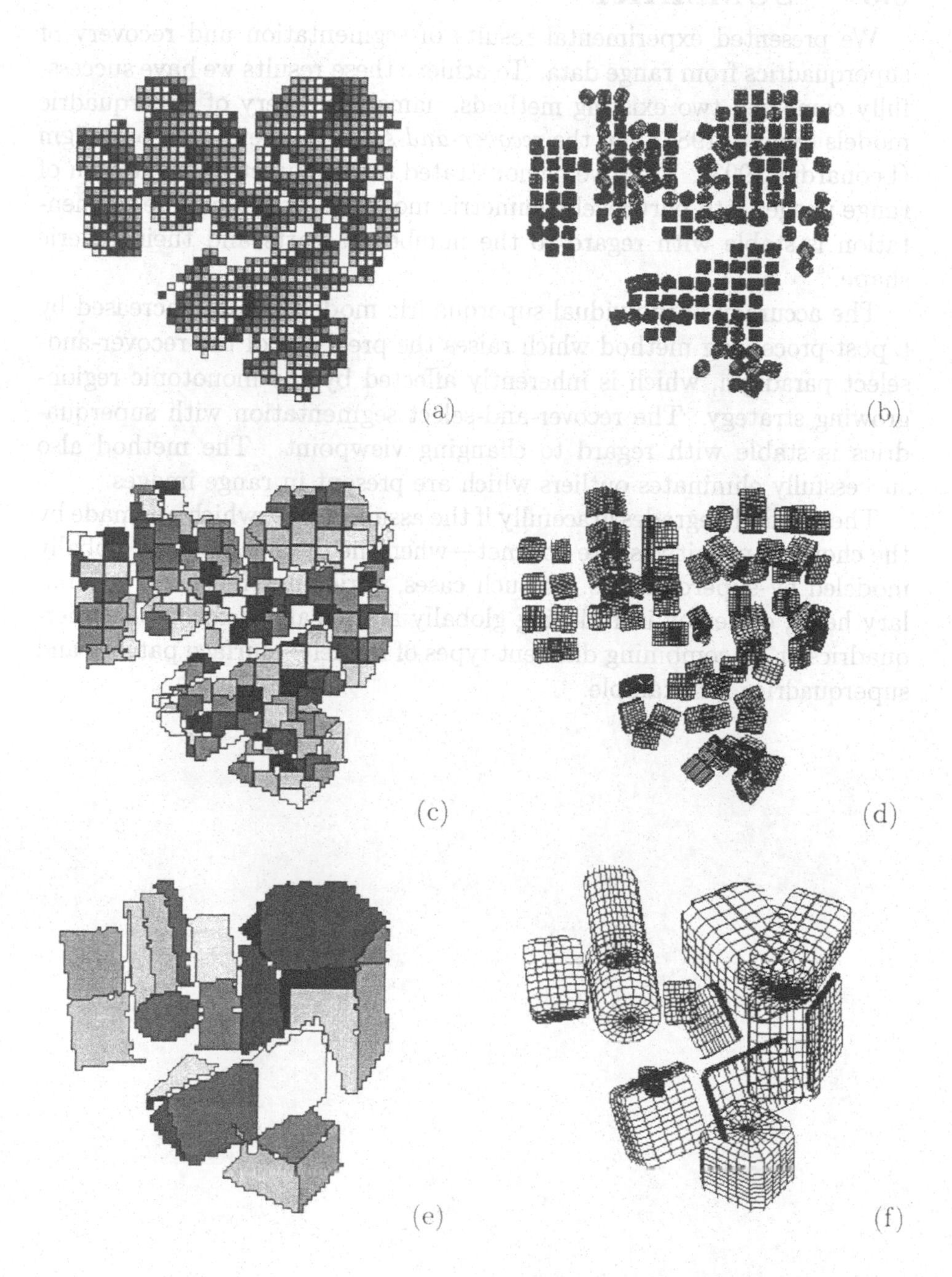

Figure 6.20. Segmentation stages for volumetric and surface models: (a), (c), (e) segmentation stages for surface models, (b), (d), (f) segmentation stages for superquadric models.

6.3 SUMMARY

We presented experimental results of segmentation and recovery of superquadrics from range data. To achieve these results we have successfully combined two existing methods, namely recovery of superquadric models (Solina, 1987) and the *recover-and-select* segmentation paradigm (Leonardis, 1993). Thus we demonstrated that a direct segmentation of range images into part-level volumetric models is possible. The segmentation is stable with regard to the number of parts and their generic shape.

The accuracy of individual superquadric models can be increased by a post-processing method which raises the precision of the recover-and-select paradigm, which is inherently affected by the monotonic region-growing strategy. The recover-and-select segmentation with superquadrics is stable with regard to changing viewpoint. The method also successfully eliminates outliers which are present in range images.

The method degrades gracefully if the assumptions, which are made by the choice of primitives, are not met—when the data can not be globally modeled by superquadrics. In such cases, enriching the shape vocabulary helps, either by introducing globally and locally deformable superquadrics or by combining different types of models—surface patches and superquadrics for example.

Chapter 7

APPLICATIONS OF SUPERQUADRICS

The earlier chapters have already shed some light on the utility of superquadrics in computer vision. The primary role of superquadrics has been in filling a much felt need for a volumetric representation in conjunction with shape recovery and segmentation. Despite initial reluctance in using superquadrics due to their nonlinear form, they have proven to be the modeling primitives of choice for many applications that require volumetric models. It is important to understand the limitations of the basic superquadric models, namely their limited shape vocabulary, and the fact that they are really coarse-grain closed volumetric models suitable, in particular, for object-centered generic shape descriptions. Like any other shape primitive, superquadrics can not represent arbitrary shapes. In the power of representing shapes, they are a subset of generalized cylinders, with a non-linearly varying superelliptical cross-section. For example, complex shapes such as bifurcating elongated objects (blood vessels) are better described by a generalized cylinder defined by a spine function and a cross-section function.

When selecting a volumetric representation for a vision application, one has to keep in mind the facts that superquadrics have nonlinear relation among parameters, a smooth surface, no real existence of edges, and inherent global symmetry assumptions which make them insensitive to small occlusions and change of viewpoint. Some of the deficiencies in the flexibility of the superquadric shapes can be overcome by attaching to the superquadric a surface mesh with local control, or by blending (see Chapter 3).

Some researchers who have experimented with superquadrics switched later to more specialized models which better suited their needs (Boult and Gross, 1988; Bobick and Bolles, 1992; Delingette et al., 1993).

Others have augmented the superquadrics to model locally deformable shapes and defined other recovery formulations which are physics-based and hence, more general in their applications (Pentland and Williams, 1989a; Terzopoulos and Metaxas, 1991; Gupta and Liang, 1993).

In addition to applications in computer vision and computer graphics, superquadrics are also used in robotics for representing the manipulated objects or the workspace of the manipulator kinematic chain itself. The first part of the chapter is therefore, an overview of various applications of superquadrics in computer vision, computer graphics, and robotics. The second part of the chapter shows how we have used superquadrics for registration of range images.

7.1 A SURVEY OF SUPERQUADRIC APPLICATIONS

7.1.1 VISION APPLICATIONS

Superquadrics are used in computer vision applications as volumetric shape primitives for 3D representation, segmentation, object classification, recognition, and tracking. Superquadrics are good at capturing the global coarse shape of 3D objects or, of their constituent parts. The addition of global deformations increases the expressive power of superquadrics, but still limits it to the global coarse shape as opposed to local details. This lack of fine scale representation can be addressed by adding local degrees of freedom. However, one drawback of such locally deformable extensions is that they have too many degrees of freedom to meaningfully segment even a simple scene. The increase in expressive power also results in an increase in complexity of all the visual tasks like scene segmentation, object representation, recognition, and object classification. Fig. 7.1 shows various aspects of computer vision applications as a function of model complexity. Consequently, most of the segmentation and classification work (Pentland, 1990; Gupta and Bajcsy, 1993; Ferrie et al., 1993; Leonardis et al., 1997) is restricted to non-deformable or only globally deformable superquadric models, while the role of local deformations is reserved for better description of surface details. Therefore, we concern ourselves here only with the various aspects of the globally deformable superquadrics.

The earliest works on superquadrics dealt primarily with single model analysis, since segmentation of complex scenes required model recovery to be understood first. These methods focussed either on classification of single models, where the power of superquadrics as a compact parametric model was exploited (Solina and Bajcsy, 1989; Horikoshi and Kasahara, 1990; Raja and Jain, 1992), or on using superquadrics as a vol-

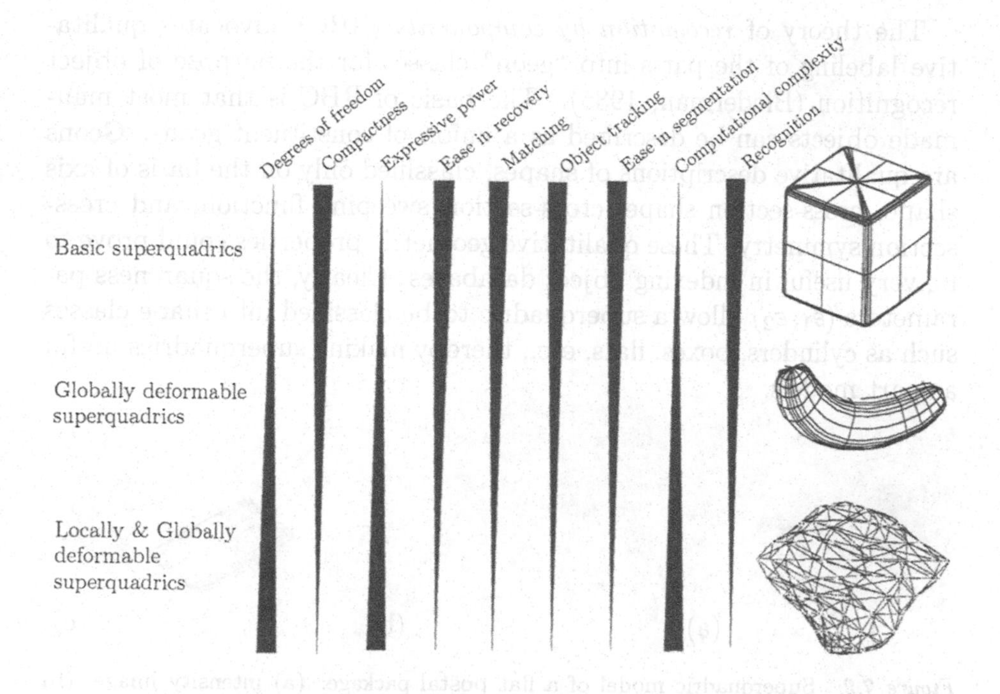

Figure 7.1. Model expressiveness as a function of complexity of addressing vision problems

umetric primitive *after* a segmentation had been obtained (Gupta et al., 1989b; Ferrie et al., 1993; Darrell et al., 1990). Once the model recovery was understood, more sophisticated techniques were designed to apply superquadrics to scene segmentation (Pentland, 1987; Dickinson et al., 1992a; Gupta and Bajcsy, 1993; Horikoshi and Suzuki, 1993; Raja and Jain, 1994; Leonardis et al., 1997), and to active exploration (Whaite and Ferrie, 1991; Ferrie et al., 1993; Whaite and Ferrie, 1997).

7.1.1.1 OBJECT CATEGORIZATION AND CLASSIFICATION

One of the important tasks in computer vision lies in classifying the observed object or its parts in broad shape categories or classes. One advantage of categorization is that it is qualitative in nature and perceptually intuitive. Thus "a cylinder" and "a block" will be partitioned into separate classes due to the differences in overall shape features. This coarse partitioning can be refined to discriminate between a tapered and a non-tapered cylinder, though at the expense of increased computational complexity and ambiguity in description (a non-tapered cylinder can appear as a tapered cylinder under perspective projection).

The theory of *recognition by components* (RBC) advocates qualitative labeling of the parts into "geon" classes for the purpose of object recognition (Biederman, 1985). The basis of RBC is that most man-made objects can be described as a union of constituent geons. Geons are qualitative descriptions of shapes, classified only on the basis of axis shape, cross-section shape, cross-section sweeping function, and cross-section symmetry. These qualitative geometric properties could prove to be very useful in indexing object databases. Clearly, the squareness parameters $(\varepsilon_1, \varepsilon_2)$ allow a superquadric to be classified into shape classes such as cylinders, boxes, flats, etc., thereby making superquadrics useful as part models.

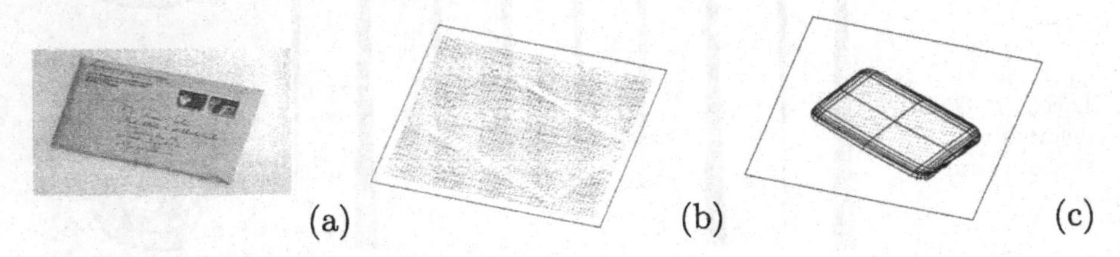

Figure 7.2. Superquadric model of a flat postal package: (a) intensity image, (b) range image, (c) superquadric model

Since most postal packages can be represented by a single superquadric model, postal objects can be described with globally deformable superquadrics and categorized as flats (Fig. 7.2), tubes, parcels, and irregular packages based on the shape and size parameters of the recovered superquadric models (Solina and Bajcsy, 1987; Solina and Bajcsy, 1989). Rules for categorization of mail pieces which correspond to actual categorization used in sorting of postal objects are in Table 7.1.

This demonstration of the feasibility of describing individual postal objects using superquadrics was extended to cluttered scenes first, by segmenting the range image using an independent edge-based scheme and, next, recovering the superquadric models of individual postal objects with the additional constraint of the physical support plane (Gupta et al., 1989b). The method was tested on hundreds of images from different types of range scanners. These classification approaches constrained the superquadric shape parameters ε_1 and ε_2 to lie between 0.1 and 1.0.

The classification scheme based on the squareness parameter was extended by partitioning the superquadric parametric space between $0 < \varepsilon_1 \leq 2.0$ and $1.0 \leq \varepsilon_2 \leq 3.0$ to develop a shape indexing language (Horikoshi and Kasahara, 1990). This representation space was mapped to verbal instructions like "rounder", "pinch", "flatten", "swell", etc., and a man-machine interface to construct object models was developed.

Table 7.1. Classification rules for postal packages

input: Superquadric parameters: a_1, a_2, a_3, ε_1, ε_2
Size parameters for postal packages:
Flat $_{\text{THICKNESS}}$, Flat $_{\text{WIDTH}}$,
Box $_{\text{SIZE}}$,
Roll $_{\text{DIAMETER}}$, Roll $_{\text{LENGTH}}$

if (($a_1 <$ Flat $_{\text{THICKNESS}}$ **and** a_2, $a_3 >$ Flat $_{\text{WIDTH}}$) **or**
($a_2 <$ Flat $_{\text{THICKNESS}}$ **and** a_1, $a_3 >$ Flat $_{\text{WIDTH}}$) **or**
($a_3 <$ Flat $_{\text{THICKNESS}}$ **and** a_1, $a_2 >$ Flat $_{\text{WIDTH}}$))
then mail package is a FLAT

else if ($a_1 >$ Box $_{\text{SIZE}}$ **and** $a_2 >$ Box $_{\text{SIZE}}$ **and** $a_3 >$ Box $_{\text{SIZE}}$)
and $\varepsilon_1 < 0.5$ **and** $\varepsilon_2 < 0.5$
then mail package is a BOX

else if ($a_1 >$ Roll $_{\text{DIAMETER}}$ **and** $a_2 >$ Roll $_{\text{DIAMETER}}$ **and**
$a_3 >$ Roll $_{\text{LENGTH}}$) **and** $\varepsilon_1 < 0.5$ **and** $\varepsilon_2 > 0.5$
then mail package is a ROLL

else mail package is an IRREGULAR PACKAGE

The authors also described an indexing scheme where complex objects were stored as superquadric models and indexed by model parameters.

Deformable superquadrics were selected also for the symbol-level representation in a 3D shape ontology (Tijerino et al., 1994). The proposed ontology for 3D visual knowledge consists of two levels:

1. a *knowledge-level*, which represents in verbal form 3D visual concepts (basic shapes, deformations of these shapes, states and features of the shapes), and

2. a *symbol-level*, which provides the operationalization of shape concepts and modifications of those concepts.

The goal of this research was to communicate user intentions to the computer with natural language and simple hand gestures to generate, manipulate and modify 3D shapes in real time. This type of interaction was named "What You Say Is What You See". Computer-supported cooperative workspaces based on virtual reality environments and other applications are candidates for such type of interactions.

Obtaining consistent (canonical) descriptions from image data in terms of superquadric part-models remains an open problem. Superquadrics can be used also for classification of shapes into geon-like classes from intensity images (Dickinson et al., 1992a) and range images (Raja and Jain, 1994; Wu and Levine, 1994a).

Dickinson chose for representation of parts, a set of ten superquadric primitives (Dickinson et al., 1992a; Dickinson et al., 1992b; Metaxas and Dickinson, 1993). The knowledge about primitives is constructed in a top-down fashion by generating a hierarchical *aspect* representation based on the projected surfaces of the primitives, with conditional probabilities capturing the ambiguity of mappings between levels of the hierarchy. The system works by grouping the segmented image regions into aspects, from which volumetric primitives are inferred. Thus, their system provides an automatic segmentation and mapping of a 2-D viewer-centered data into 3-D object-centered primitives and their connectivity. This work is among the most comprehensive in multi-part primitive recovery in a purely bottom-up (data-driven) fashion from intensity data.

Raja and Jain recovered superquadrics on pre-segmented range data and mapped the superquadric shapes into 12 shape classes corresponding to a "collapsed" set of 36 different geons (Raja and Jain, 1992; Raja and Jain, 1994). Since superquadrics are a proper subset of the geon shapes, they are inadequate for modeling the whole set of geons. They defined a five element feature vector which was computed from the recovered superquadric shape and deformation parameters and used the feature vector for classification into 12 geon classes, using binary tree and k-nearest-neighbor classifiers (Raja and Jain, 1992). Raja and Jain's experiments showed about 80% reliability in mapping low-noise real range image objects into their geon counterparts.

Wu and Levine proposed seven parametric geons derived from the superquadric model (Wu and Levine, 1994a; Wu and Levine, 1994b). They fit all seven geon models to range data which is merged from several views and select the geon model with the lowest residual. The objective function consists of two terms: a distance and a normal measure. The minimization procedure is a stochastic technique, based on a fast simulated annealing method.

Pilu and Fisher developed a method of detection and recognition of qualitative parts from 2D intensity images of real objects (Pilu and Fisher, 1996b). Inspired by the projected visible contour of superquadrics, they defined a geon-like parametrically deformable contour model (PDCM). Given some manual initialization of PDCMs in intensity images, which in effect segments the data into parts, PDCMs are recovered by minimizing an objective function which corresponds to the distance between the contour model and the image edge points. Simulated annealing is used for minimization since the objective function is very irregular with many shallow minima.

7.1.1.2 OBJECT RECOGNITION

The knowledge about objects in terms of stored models with superquadric as part-primitives has not, so far, been exploited, despite the compact representation which may facilitate fast database matching. One way to avoid the problem of uniqueness is to store multiple representations of the object. But this may defeat the purpose of using a compact representation, which should be efficient and sublinear in database matching. Thus, the model database must play a role during part-segmentation, and resolve ambiguities "on the fly".

One of the reasons that superquadrics have not been used much for object recognition is the difficulty of matching superquadrics recovered from image data with superquadrics in the model data base. Data-driven bottom-up strategy of fitting superquadrics to image data does not assure the recovery of unique "canonical" superquadric models. To solve the "recognition problem" one has to match the recovered superquadric parts, which have a continuum of shapes, to stored part-models which are also based on superquadric models. This part-matching can also be viewed as a "knowledge-driven" top-down process.

The problem of matching recovered superquadrics with a set of superquadrics in a model database was analyzed by (Chen et al., 1997). The first difficulty with matching is the non-unique parameterization of the same shapes which is a result of different possible orientations of the local superquadric-centered coordinate system. For a parallelepiped ($\varepsilon_1 = 0.1, \varepsilon_2 = 0.1$), for example, the axis z can point into any one of three possible orientations. The second difficulty is that superquadric parameters $\mathbf{\Lambda}$ have a non-linear relationship. The contribution of each model parameter to the final shape depends also on the values of other superquadric parameters. If the values of a_1, a_2, and a_3 are all equal, then just by changing the value of parameter ε_1 from 0.1 to 1, a cylinder changes to a sphere. If on the other hand, we have a very thin cylinder, the same change does not affect very much the overall generic shape (Fig. 7.3). Therefore, simple comparisons based on the Euclidean distance between parameter vectors $(a_1, a_2, a_3, \varepsilon_1, \varepsilon_2, \ldots)$ which assume that each component is equally important will not work well, even if the different subspaces are rescaled. Chen therefore proposed a similarity measure for superquadrics which is based on the volume difference of two superquadrics (Chen et al., 1997). They show that the volume difference is a valid metric. Experiments confirm that the volume differ-

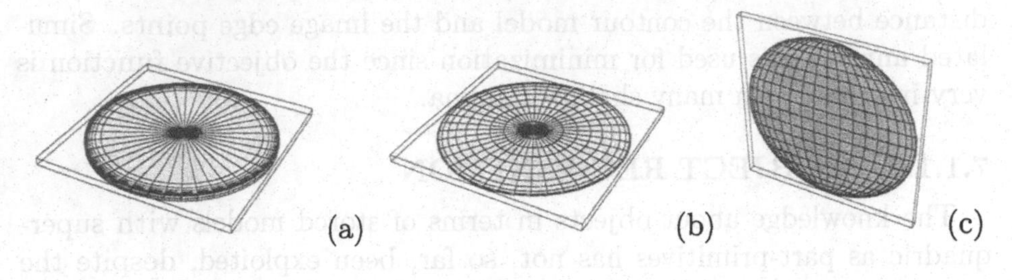

superquadric parameters	(a)	(b)	(c)
ε_1	0.1	1	1
ε_2	1	1	1
a_1	50	50	50
a_2	50	50	2
a_3	2	2	50

Figure 7.3. Superquadric models with very different sets of parameter values can have a very similar generic shape.

ence, as a similarity measure for superquadrics, enables a high success rate of recognition, except for bent superquadrics.

The problem of part-matching is in some sense, more difficult than matching completely stored models, because basic parts can combine arbitrarily to form complex shapes. Part-matching requires a piecewise, but not necessarily unique, segmentation of complex shapes into part primitives. Object recognition problems do not typically confront with object-by-parts problems. It is easy to see that object recognition based on superquadrics would essentially use the part configuration constraint from the stored models, and solve the ambiguity problem arising in arriving at the piecewise description. This is still an open problem, since it requires interleaving of database search with multiple model recovery.

The modal representation can also be used for object recognition (Pentland and Sclaroff, 1991; Sclaroff and Pentland, 1995). Similar to Fourier decomposition, the low-order modes are relatively unaffected by local shape variations and by measurement noise. A modal representation reduced to low-frequency modes offers a unique representation of shape which can be effectively used for recognition, comparison, and other database tasks.

A framework for parametric shape recognition using superquadrics and based on a probabilistic inverse theory was proposed (Arbel et al., 1994). Conditional probability density functions which are needed in this framework can be automatically generated. The recognition process takes into account the *a priori* knowledge of the objects comprising the database, as well as the information obtained from the process of superquadric model estimation. A feedback from the recognition task is incorporated similar to the feedback from the model building level proposed in the autonomous exploration framework (Whaite and Ferrie, 1997).

Superquadrics are also part of a two-stage hybrid model for object recognition (Ayoung-Chee et al., 1996). Superquadrics are used in the first stage for a stable recovery of the overall shape of objects and to provide a convenient abstraction mechanism for robust indexing. The second, finer level consists of surface patches of uniform mean and Gaussian curvature which model the error in superquadric fit to the data. These two stages or levels of the hybrid model are therefore not independent, but supplement each other.

Superquadrics were tested on classification of electronic components such as electrolytic capacitors, integrated circuits, and crystals mounted on printed circuit boards (van Dop and Regtien, 1998). For superquadric recovery of pre-segmented range data, the authors have used a gradient least-squares minimization of an error-of-fit measure which was based on the radial distance (Whaite and Ferrie, 1991), and augmented with a background constraint and a robust technique for elimination of outliers. Out of 144 objects in the database, 123 were correctly classified.

For a more general form of recognition, a system must recognize entire classes of particular types of objects. For example, all possible chairs or all possible dogs. Due to extreme variations in structure and shape of individual instances of the same object class, such as chairs, there is no ideal shape prototype from which all other instances could be derived by simply changing the values of some parameters. Therefore, researchers in artificial intelligence (Winston et al., 1983) have early on suggested that function could be used as a means to represent and recognize generic classes of objects. Recognition of generic classes hence, requires inferring function from shape and structure (Stark and Bowyer, 1991). Due to their shape flexibility, superquadrics were soon considered as part models in such generic category models (Solina and Bajcsy, 1986).

The part-based shape recognition based on superquadrics was extended to function recognition (Rivlin et al., 1994). The authors proposed a two-level representation—a shape and a functional level which have many-to-one relations between functional and shape primitives. The system supports bottom-up recognition by recovering part shapes followed by identification based on the functionality of its parts and top-down recognition where the expected functionality of the model helps to constrain the search for shape features in the image.

Such static analysis of function, based purely on shape, can not verify elements of function that depend on properties other than shape. Interaction with the object is needed, for example, poking and handling to test the stability and material properties of the object (Stark, 1994). Stark formulated an approach to dynamic analysis of function based shape in which the results of an initial static analysis of shape direct a plan of interacting with an object to verify that it has the appropriate physical properties. Stark used the *ThingWorld* modeling system (Pentland, 1989b) based on superquadric models to reason about solid shapes and physical properties of objects.

7.1.1.3 3D REPRESENTATION USING SUPERQUADRICS

In addition to the vision objectives of object categorization, segmentation, and recognition, superquadrics have been used for intermediate and final 3D representations for various purposes. Superquadrics can play a useful role in building a range image representation, registering range images, merging range images or their representations, and guiding next-viewpoint planning.

Superquadrics are popular for modeling the heart since its shape and the shape of its constituent parts can be easily approximated with deformed superquadrics. The global shape and motion of a superquadric can be decoupled from local degrees of freedom of the deformable superquadric (Chen et al., 1994a). The global superquadric model accounted for the global degrees of freedom of a beating heart by tracking the segmented left ventricular data obtained from a CT scanner. The local degrees of freedom were accounted for by a spherical harmonics model. The modal analysis technique was also used for the description of a beating heart from the 2-D outline of the left ventricle in X-ray images (Pentland et al., 1991). Modeling of the myocardium with locally deformable superquadrics was proposed also by others (Young and Axel, 1992; McInerney and Terzopoulos, 1993; Park et al., 1994; Bardinet et al., 1994; Gupta et al., 1994; O'Donnell et al., 1995).

Another anatomical shape that can be modeled with superquadrics is the cranium. Researchers used superquadrics to quantify the cranium surface asymmetry (Shiang et al., 1993). Cranial asymmetry which is a result of asymmetry of the human brain is explained with cerebral laterization. Cerebral lateral dominance is related to several diverse phenomena. The best known phenomena is the localization of speech to the left hemisphere in the majority of humans. Instead of traditional anthropometric measurements using sliding calipers between pairs of replicable landmarks on the skull, superquadrics were fitted to more than 200 random 3D points on the skull. Asymmetry was quantified as the radial deviation of the surface data points from their symmetric best-fit superquadric models. For fitting superquadrics, the authors used our problem formulation (Solina and Bajcsy, 1990). Superquadrics were also used to document rapid change of the cranium form in response to environmental influences (Ripley et al., 1995).

Globally deformable superquadrics, enhanced with several levels of local deformations, were used to model parts of the human body (Metaxas et al., 1997). The method enables shape representation at various levels of abstraction.

Superquadrics were used in a prototypical CAD system for the garment industry (Jojić and Huang, 2000). The goal of the system was to model individual customers with superquadrics (Fig. 7.4). Superquadrics, enhanced by local deformations, were extracted from 2D images and stereo, aided by structured light.

Superquadrics have been used for modeling the boundaries of human motion to study movement disabilities (Marzani et al., 1997). Three cameras are used to acquire synchronized sequences of a human in motion. A human leg can be represented with three tapered superquadrics. From each view, 2D superquadric shapes with circular cross-section are reconstructed first, and then merged into tapered superquadrics, to get an articulated 3D model of a leg in a sequence of images. For reconstruction of 2D tapered superquadrics, the authors used an iterative fuzzy clustering method.

Superquadrics have been used as coarse-level models for modeling outdoor scenes (Pentland, 1986; Bobick and Bolles, 1992), where only an approximate shape and size of the object is of interest. Superquadrics are used in a "real-time" physics-based tool for solid modeling (Pentland, 1989b). This *ThingWorld* modeler has an intuitive solid modeling design interface which makes it easy for naïve users to quickly create, modify, and interact with complex designs.

We have also addressed the problem of reverse engineering by constructing CAD models from range images where superquadrics can be used as an intermediate coarse model for segmentation and registration of range images (Jaklič and Solina, 1994; Solina et al., 1998; Martin and Várady, 1998). The basic objective of this project was to construct boundary representation solid models of three-dimensional objects from measured data. Input data can be acquired from a variety of sources, mainly laser scanners of different resolutions and accuracy. The aim was to construct a conventional boundary representation CAD model in as automatic a manner as possible and to cope with a wide variety of object types, ranging from simple objects bounded by planar, cylindric and

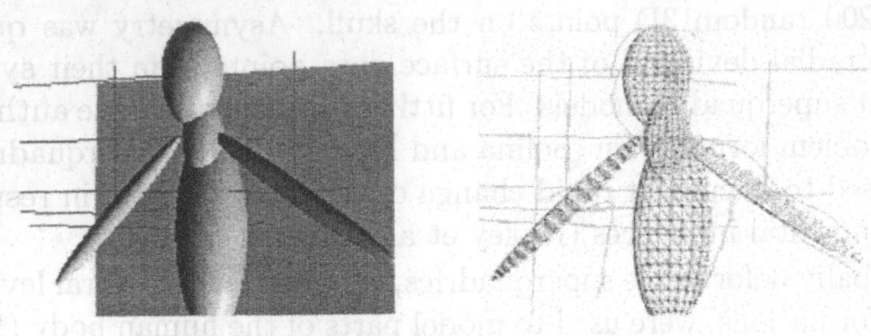

Initial position of superquadric models

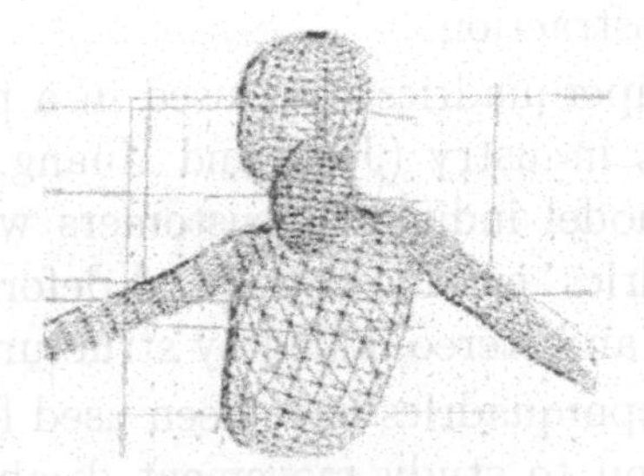

The final reconstructed locally deformed superquadrics

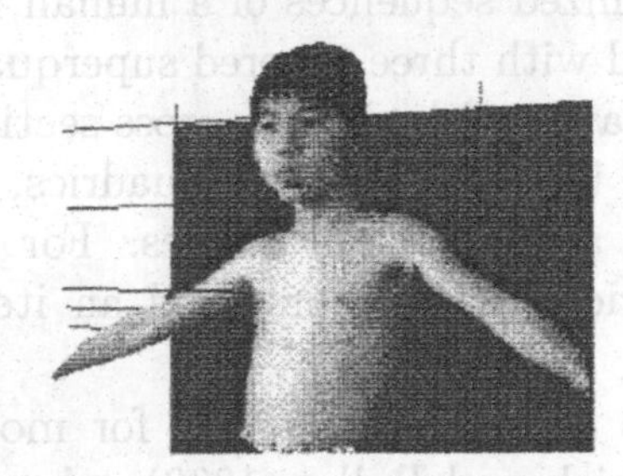

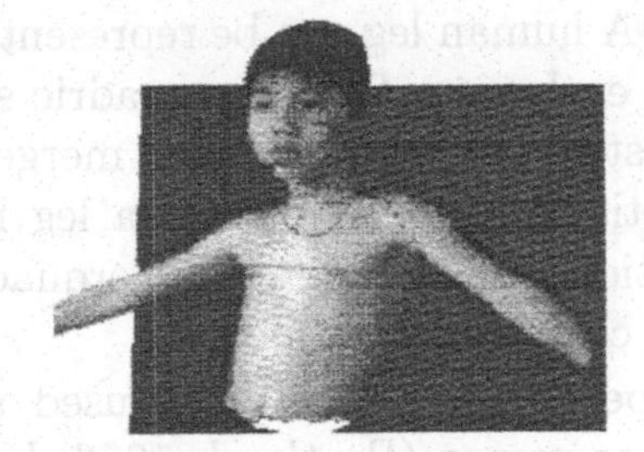

Texture mapped reconstructed surface

Figure 7.4. Superquadric-based CAD system for modeling human bodies (Jojić and Huang, 2000). The figure is reproduced with the permission of the authors.

spherical faces, and simple objects with blends, up to general free-form objects bounded by arbitrary curved surfaces.

7.1.1.4 IMAGE COMPRESSION

Upcoming standards for video coding (MPEG-4 and MPEG-7) include the possibility of segmenting the scene in different separate entities before coding which leads to a more flexible and intelligent coding. Every part can have its own coding algorithm and a different coding update frequency which is best suited to the importance and visual characteristics or motion of the segment. The segmentation information must be encoded with a small number of bits. The borders of such regions or objects can be approximated with superellipses (Wuyts and Eycken, 1997) as shown in Fig. 7.5. Besides a very compact description of shape, superellipses also offer the possibility to simply evaluate a cost function needed for segmentation which represents the preference for a pixel to belong to an object. If the shape of the object is complex, additional superellipses can be added or subtracted from the existing ones.

(a)
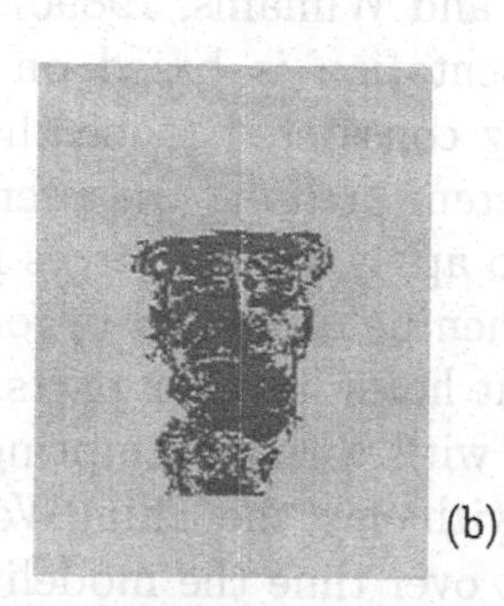
(b)
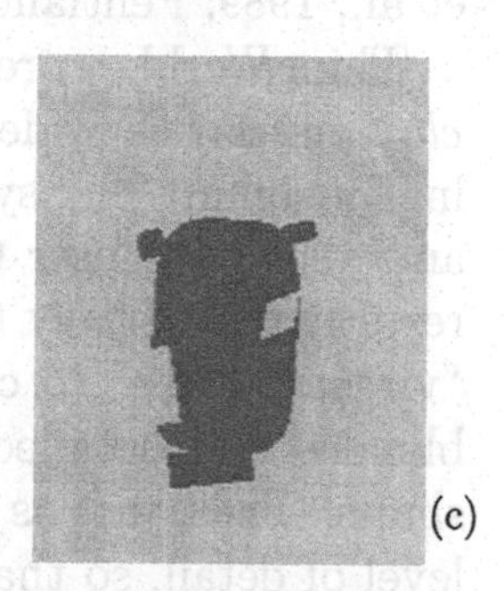
(c)

Figure 7.5. Superquadric based encoding of segments for object-based video coding (Wuyts and Eycken, 1997): (a) original image, (b) segmented region, (c) approximation of the region with six superellipses (2D equivalents of superquadrics) which requires only 6×33 bits for encoding. From the superellipse which approximates the whole face, two smaller ones are subtracted, and three are added.

7.1.2 GRAPHICS APPLICATIONS

A major benefit of superquadrics in computer graphics is their double formulation: the interchanging implicit and explicit equations. In this way, a large range of visualization algorithms can be used. Projective techniques would generally use the explicit or parametric formulation which enables easy tesselation, whereas ray tracing techniques prefer the implicit one which can be used for testing the ray intersections.

Most commercial computer graphics software packages nowadays include superquadrics as one of several possible 3D building blocks. In this way, superquadrics are used in many different modeling applications. A custom built virtual environment, which is used for pilot training, uses superquadrics for rendering shapes of man-made and natural objects (McCarthy et al., 1993). Some issues related to efficient rendering of superquadrics which is of essential importance to computer graphics applications are discussed in Chapter 2.

7.1.2.1 THINGWORLD MODELING SYSTEM

Based on superquadric models, Pentland and his group developed the *ThingWorld* modeling system (Pentland, 1989b) with a dual purpose:

first, to provide a solid modeling design interface which is intuitive — making it easy for naïve users to quickly create, modify, and interact with complex designs — and second, to provide an artificial intelligence and machine vision test environment which is able to reason about solid shape and the physical properties of objects, and which can automatically recover CAD-like models from image or range sensor data (Pentland et al., 1989; Pentland and Williams, 1989c).

ThingWorld representation is based on the notion of *sculpting in clay*, as clay is widely considered to be the fastest traditional modeling medium. The system presents the user with virtual lumps of clay and allows the user to apply virtual forces in order to form them. The resulting parts may then be combined to form larger objects or used as "virtual knives" to cut holes in other parts. The surfaces may then be blended and detailed with various sculpting tools to produce the final shape. Thus, just as with clay, the ThingWorld system allows a variable level of detail, so that over time the modeling progresses from rough to fine forming operations.

Perhaps the most novel aspects of the ThingWorld system are its use of a physical, force-based interface in a real-time modeling framework, its ability to vary the level of detail being modeled, and the ability to automatically generate models from measurements of real objects. These abilities stem primarily from the use of a modal representation of object dynamics and deformation rather than the more common finite element representation. Use of the modal method provides both a large speed-up in the calculation of the deforming object dynamics and a way of progressively adjusting the level of modeling details.

7.1.3 ROBOTICS APPLICATIONS

Superquadrics are used in robotics for modeling the objects of manipulation which is required for planning of grasping, the robot workspace and the robotic manipulator itself.

7.1.3.1 GRASPING

Grasping can be used for manipulation by picking up objects, as well as for tactile recognition. Superquadrics were used for obtaining initial global estimates of object's gross contour and volume as part of a larger system for 3D shape recovery and object recognition using touch and vision methods (Allen and Michelman, 1990). For grasping by containment, they used the Utah-MIT hand equipped with tactile sensors which provided a fair amount of sparse point contact data. These points were the input for our superquadric recovery algorithm (Solina and Bajcsy, 1990). Although there were far fewer touch points (approximately 60) than range points in a typical range image of a comparable object, the superquadric recovery performed well.

What makes superquadrics particularly relevant for haptic recognition is that the models are volumetric in nature, which maps directly into the psychophysical perception processes suggested by grasping by containment.

Superquadrics were used also for grasp planning (Ikeuchi and Hebert, 1996). The robot gripper mounted on the Mars Rover, which was developed at the Carnegie-Mellon University, had the task of picking up rock samples from the ground. In the range image of the area in front of the Rover robotic arm, areas corresponding to larger objects on the flat ground were isolated with a simple segmentation method. The range data of each of these isolated regions was the input data to our superquadric recovery method (Solina and Bajcsy, 1990). The recovered superquadrics were used for guiding the robot gripper to pick up the objects.

7.1.3.2 PATH PLANNING

Superquadrics can be used for modeling objects that the robot is manipulating with or the whole mobile robot work environment (Jouvencel and Simphor, 1991), as well as for modeling the robot kinematic chain itself (Agba et al., 1993). For control of kinematic chains such as robot manipulators and legged robots, simulation of interactions of kinematic links with objects in the environment is required. Superquadric models are used for modeling individual links of the manipulator which are related by homogeneous transformations. Using the inside-outside function, it is possible to determine distinctly whether an arbitrary point falls inside or outside of the volumes defined by the superquadrics. This modeling technique has been implemented in a hybrid simulator for undersea telerobotic manipulation for collision detection and grasp planning (Agba et al., 1993).

Khatib, who introduced the concept of artificial potential fields for path planning, proposed superquadric as objects models in the artificial potential field (Khatib, 1986). Artificial potential technique surrounds the obstacles with repulsive potential energy functions and places the goal in an attractive well. Using the potential field, manipulators and mobile robots can reach a desired destination without collision with obstacles.

Volpe and Khosla continued the work with superquadric potential functions for modeling obstacles and goal positions in robot work space (Volpe and Khosla, 1990). Local methods for path planning which are suitable for real-time control express these artificial potential functions in the Cartesian workspace of the manipulator. Adding attractive and repulsive potentials can create local minima in the potential function which is not desirable for robust control. Since superquadric potential functions are asymptotically spherical in nature, these spherical surfaces prevent the formation of local minima in the potential function so that obstacles can be successfully avoided. On the other hand, isopotential surfaces near the object may closely model the object. The approach potential can hence provide deceleration forces that ensure a safe contact velocity at the surface of the goal position. Superquadric potential functions were applied in a newly developed robot controller in an experimental setting. Visual feedback was also planned that could provide obstacle data in real-time, enabling dynamic obstacle avoidance. Providing that the vision system also employs superquadric models, the integration of visual feedback into the robot control would not be difficult. However, a word of caution is in place here—the isopotential functions based on the superquadric implicit function, equation (2.15) on page 21, are not equidistant functions (see also Fig. 4.4 where superquadric isopotential functions are shown).

7.2 RANGE IMAGE REGISTRATION

In this section we elaborate on the use of superquadrics for range image registration (Jaklič, 1997). It is generally not possible to recover a geometric precise superquadric or any other description of an object from a single range image, even though the data presented in the range image is modeled well. This is due to self-occlusion of the object, where one part occludes another part, and also due to a measuring principle like triangulation which structured light range scanners use (Maver and Bajcsy, 1993). Besides, recovery of superquadrics from a single range image is underconstrained and even the additional constraint of minimal volume which we use, does not guarantee a perfectly precise model for a single superquadric-like object (Whaite and Ferrie, 1991).

Thus, to form a geometric precise model, range images of an object must be acquired from different viewpoints and merged together into a common coordinate frame. Range image registration is the process of finding a rigid transformation between the range images. The transformations are then used to merge the data. Many registration algorithms are based on some form of local minimization and require a good initial estimate of the transformation (Chen and Medioni, 1992; Besl and McKay, 1992; Turk and Levoy, 1994; Blais and Levine, 1995; Dorai et al., 1996; Eggert et al., 1996).

Initial estimate of transformation can be provided by image feature correspondence (Bergevin et al., 1995), a range sensor (Blais and Levine,

1995), an object manipulation device (Soucy and Laurendeau, 1995) and even by a human operator (Turk and Levoy, 1994; Dorai et al., 1996). In this section, we focus on the most challenging way of providing the initial estimate from image features correspondence, which is also an implicit mechanism used by a human operator. The additional motivation to explore this approach is a possible extension of these methods for object pose estimation and object recognition. The basic requirement for a feature correspondence method is that the method should not be too sensitive to differences in computed features caused by a changing viewpoint.

One commonly used method to determine the transformation is to attach a coordinate frame to each range image based on object's principal axes of moments of inertia and object's center of gravity. The moments and center of gravity are calculated directly from the range image data points. The direct use of data points is susceptible to two problems. First, the density of points for a surface patch is dependent on the relative orientation of the surface normal with respect to the scanning direction, and second, self-occlusion might cause that different parts of the object surface are used in calculation of moments. This effectively reduces the use of method to the cases where rotation of the object is small.

A method that would use volumetric superquadric models recovered from a single range image should not suffer from the first problem, while the influence of the second one should be reduced, but not eliminated, since some types of self-occlusion can prevent the correct inference of the volumetric model. For example, three sides of a block are always visible in a non-degenerate view and from them we can always reconstruct a precise volumetric model of the object. The superquadrics present good volumetric model candidates since they can be reasonably well recovered from a single range image. Besides, a closed form expressions for calculation of their moments exist (see Section 2.4.7 in Chapter 2). An additional benefit of superquadrics is that they approximate a subset of polyhedral shapes, with a finite number of aspects views, as well as a subset of smooth shapes, with an infinite number of aspect views.

Liang, Lin and Chen (Liang et al., 1992) used superquadrics as an intermediate representation to establish correspondence between two 3D data sets. The model-to-data correspondence was used for model interpolation in 3D morphing applications. The model-to-data correspondence scheme is also shown to be useful for mapping highly concave shapes to convex models (Gupta and Liang, 1993).

In the following section we will apply the classical moment method to compositions of superquadrics produced by our segmentation algorithm.

7.2.1 DEFINITION OF MOMENTS

An ordinary 3D moment of an object of finite volume in a Cartesian coordinate system is defined as

$$m_{ijk} = \int_V x^i y^j z^k dV, \tag{7.1}$$

where the integer $(i + j + k)$ is the order of m_{ijk}. In the following discussion we will use moments of various orders, as well as their sums. For notational convenience and emphasis of physical interpretation of moments, we introduce the shorthand notation presented in Table 7.2. We define an object's vector $\mathbf{i}$ of first order moments and volume as

$$\mathbf{i} = \begin{bmatrix} I_x \\ I_y \\ I_z \\ V \end{bmatrix} \tag{7.2}$$

and its matrix of inertia $\mathbf{I}$ as

$$\mathbf{I} = \begin{bmatrix} I_{xx} & I_{xy} & I_{xz} \\ I_{xy} & I_{yy} & I_{yz} \\ I_{xz} & I_{yz} & I_{zz} \end{bmatrix}, \tag{7.3}$$

where

$$I_{xy} = -D_{xy}, \; I_{yz} = -D_{yz}, \; I_{xz} = -D_{xz}. \tag{7.4}$$

Table 7.2. Notation for different moments used in range image registration

Notation	*Definition*	*Physical interpretation*
V	m_{000}	volume of the object
I_x	m_{100}	x coordinate of the object's center of gravity multiplied by V
I_y	m_{010}	y coordinate of the object's center of gravity multiplied by V
I_z	m_{001}	z coordinate of the object's center of gravity multiplied by V
I_{xx}	$m_{020} + m_{002}$	object's moment of inertia about the x axis
I_{yy}	$m_{200} + m_{002}$	object's moment of inertia about the y axis
I_{zz}	$m_{200} + m_{020}$	object's moment of inertia about the z axis
D_{xy}	m_{110}	object's moment of deviation
D_{xz}	m_{101}	object's moment of deviation
D_{yz}	m_{011}	object's moment of deviation

7.2.2 TRANSFORMATION OF MOMENTS

Moments are defined with respect to a particular coordinate system. It is possible to derive relationships between moments computed in various coordinate frames from the relationships between the coordinate systems. In this subsection we will derive the relationships for vectors $\mathbf{i}$ and matrices $\mathbf{I}$. In proceeding subsections we will use these relationships to compute vector $\mathbf{i}$ and matrix $\mathbf{I}$ of a composition of superquadrics with respect to the global coordinate system from the corresponding vectors $\mathbf{i}_i^L$ and matrices $\mathbf{I}_i^L$ of individual superquadrics computed in their canonical coordinate systems. Later, we will also use these relationships to determine a set of four canonical coordinate systems, which are rigidly "attached" to the composition of superquadrics.

7.2.2.1 TRANSFORMATION OF VECTORS

We will name the two coordinate systems the *global coordinate system* G (coordinate system of the range scanner) and the *local coordinate system* L (canonical coordinate system of a superquadric). The transformation from the homogeneous coordinates $\mathbf{x}^L$ expressed in L to homogeneous coordinates $\mathbf{x}^G$ expressed in G is given by matrix $\mathbf{T}$ (see also Section 2.3 in Chapter 2)

$$\mathbf{x}^G = \mathbf{T}\mathbf{x}^L. \tag{7.5}$$

To indicate the coordinate system in which the moment is defined, we add a superscript referring to the name of the system. If there is no superscript, we assume the global coordinate system G. It is trivial to show that the zero order moment and first order moments are transformed as follows

$$\mathbf{i} = \begin{bmatrix} I_x \\ I_y \\ I_z \\ V \end{bmatrix} = \begin{bmatrix} \int_V x dV \\ \int_V y dV \\ \int_V z dV \\ \int_V dV \end{bmatrix} = \mathbf{T} \begin{bmatrix} I_x^L \\ I_y^L \\ I_z^L \\ V^L \end{bmatrix} = \mathbf{T}\mathbf{i}^L. \tag{7.6}$$

A physical interpretation of the derived relationship is that the volume of the object and the center of gravity are invariant to the selected Cartesian coordinate system.

7.2.2.2 TRANSFORMATION OF MATRICES

If we try to use homogeneous transformations to derive the relationships between moments of inertia and deviation in the two coordinate systems, this leads to complicated expressions which cannot be expressed as matrix multiplication, as is the case for zero and first order moments. Instead, we decouple the homogeneous transformation into translation and into rotation. This leads to simpler expressions for each of the transformations. The first expression, in its special form, is known as Steiner's formula and the second as a tensor transformation. It is always possible to decouple a rigid transformation into a sequence of a rotation, followed by a translation or vice versa. Note, however, that the translation and rotation are not commutative and that the corresponding translational and rotational matrices are not equal! Thus

$$\mathbf{x} = \mathbf{T}\mathbf{x}^L = \mathbf{T}_{tra}\mathbf{T}_{rot}\mathbf{x}^L = \mathbf{T}'_{rot}\mathbf{T}'_{tra}\mathbf{x}^L. \quad (7.7)$$

After the decomposition, we apply the corresponding moment transformations one after another.

Transformation of matrix I under pure translation First, we will examine how the moments of inertia are transformed under translation of the coordinate system, which is defined by a homogeneous matrix of the form

$$\mathbf{T}_{tra} = \begin{bmatrix} 1 & 0 & 0 & p_x \\ 0 & 1 & 0 & p_y \\ 0 & 0 & 1 & p_z \\ 0 & 0 & 0 & 1 \end{bmatrix}. \quad (7.8)$$

From the definition it follows that moments of inertia transform as

$$I_{zz} = \int_V (x^2 + y^2)dV = I_{zz}^L + 2p_x I_x^L + 2p_y I_y^L + (p_x^2 + p_y^2)V, \quad (7.9)$$

$$I_{yy} = \int_V (x^2 + z^2)dV = I_{yy}^L + 2p_x I_x^L + 2p_z I_z^L + (p_x^2 + p_z^2)V, \quad (7.10)$$

$$I_{xx} = \int_V (y^2 + z^2)dV = I_{xx}^L + 2p_y I_y^L + 2p_z I_z^L + (p_y^2 + p_z^2)V, \quad (7.11)$$

and deviational moments as

$$D_{xy} = \int_V xydV = D_{xy}^L + p_x I_y^L + p_y I_x^L + p_x p_y V, \quad (7.12)$$

$$D_{xz} = \int_V xzdV = D_{xz}^L + p_x I_z^L + p_z I_x^L + p_x p_z V, \quad (7.13)$$

$$D_{yz} = \int_V yzdV = D_{yz}^L + p_y I_z^L + p_z I_y^L + p_y p_z V. \quad (7.14)$$

If the origin of the local coordinate system is in the center of gravity of the object, the first order moments I_x^L, I_y^L, I_z^L are equal to 0 and the expressions for I_{xx}, I_{yy}, I_{zz} above reduce to the Steiner's law. We introduce a shorthand notation for the set of equations (7.9) to (7.14) as

$$\mathbf{I} = map(\mathbf{T}_{tra}, \mathbf{I}^L, \mathbf{i}^L). \quad (7.15)$$

Transformation of matrix I under pure rotation The homogeneous transformation matrix for pure rotation is now of the form

$$\mathbf{T}_{rot} = \begin{bmatrix} n_x & o_x & a_x & 0 \\ n_y & o_y & a_y & 0 \\ n_z & o_z & a_z & 0 \\ 0 & 0 & 0 & 1 \end{bmatrix} \tag{7.16}$$

and from the definition of matrix **I** the following relationship follows (for derivation see (Goldstein, 1980) for example)

$$\mathbf{I} = \mathbf{T}_{rot\_sub}\mathbf{I}^L\mathbf{T}^{-1}_{rot\_sub}, \tag{7.17}$$

where the $\mathbf{T}_{rot\_sub}$ is the upper left 3×3 sub-matrix of the homogeneous matrix representing pure rotation.

Composition of transformations of matrices From equation (7.7) we can write out two alternative forms for transformation of the matrix **I** under a rigid transformation of the coordinate system

$$\begin{aligned} \mathbf{I} &= map(\mathbf{T}_{tra}, \mathbf{T}_{rot\_sub}\mathbf{I}^L\mathbf{T}^{-1}_{rot\_sub}, \mathbf{T}_{rot}\mathbf{i}^L), & (7.18) \\ \mathbf{I} &= \mathbf{T}'_{rot\_sub}\; map(\mathbf{T}'_{tra}, \mathbf{I}^L, \mathbf{i}^L)\; \mathbf{T}'^{-1}_{rot\_sub}. & (7.19) \end{aligned}$$

7.2.3 FINDING AN ESTIMATE OF THE RIGID TRANSFORMATION

The basic idea of range image registration based on moments is to construct a coordinate frame, which is rigidly attached to the object in each image. After constructing the two frames, we know their relationship to the global coordinate system and thus, we also know the rigid transformation between the two frames, which is also the rigid transformation of the object. We will name the constructed frames the *canonical frames.*

A canonical frame has its origin in the center of gravity of the object. In such a frame the first order moments are equal to 0. The axes of the coordinate system are aligned along the axes of minimal and maximal moments of inertia. Both, the center of gravity, as well as the axes of minimal and maximal moment of inertia are invariant to rigid transformation of the coordinate system and are intrinsically related to the mass distribution of the object.

Since we are dealing only with the right hand Cartesian coordinate frames, we uniquely determine the remaining third axis by fixing two axes of the coordinate system. For our work, we freely selected the x and the z axes to correspond to the minimal and to the maximal moment of inertia, respectively. Note however, that the moments of inertia are invariant to rotation of the coordinate frame for 180° about any of the coordinate axes. This leads to four possible orientations of the canonical coordinate frame: the constructed one, and one for each rotation of 180° about the x, y, and z axis respectively, as shown in Fig. 7.6.

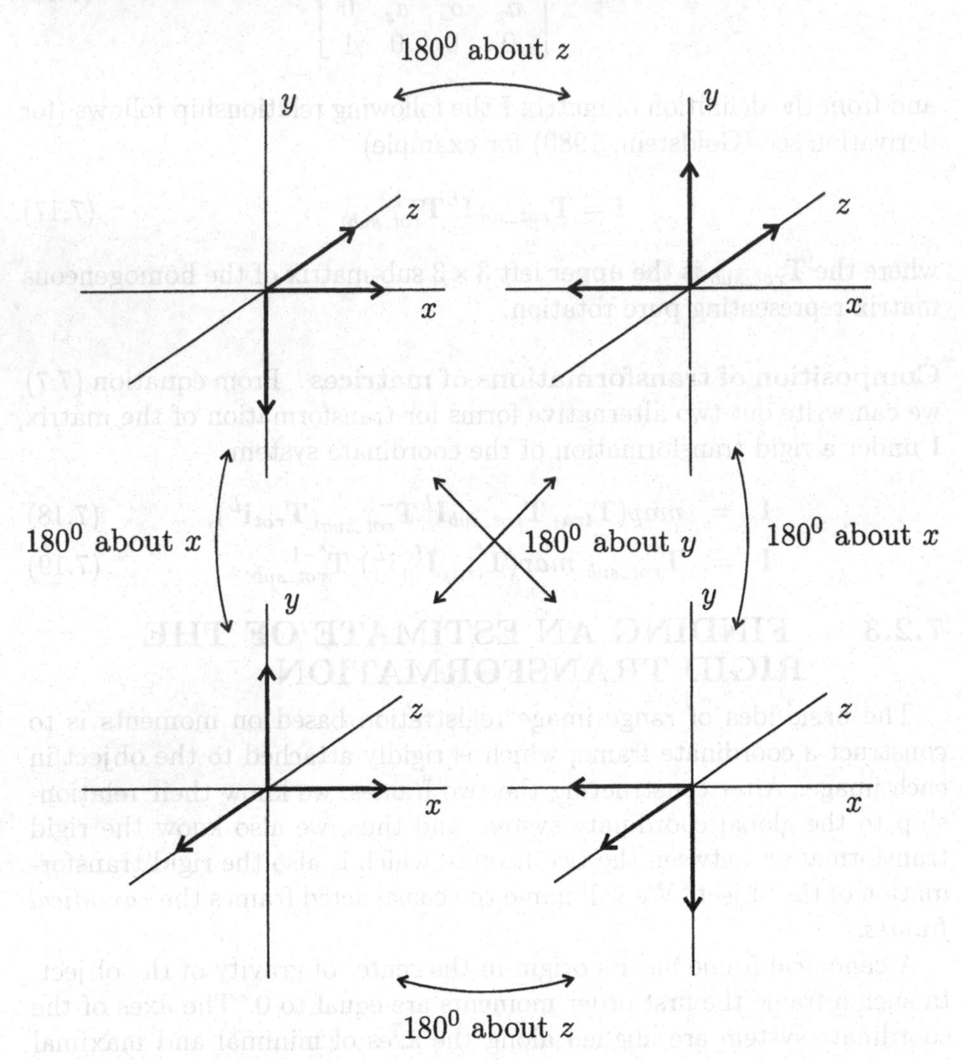

Figure 7.6. Four different right hand Cartesian coordinate frames with their axes aligned along given lines. Each one of them generates the remaining three by rotations of 180° about all the axes.

Having defined the four canonical frames, we look for the rigid transformation between the two images. In the first image we choose any of the four frames to be our canonical frame. This frame is uniquely related to the geometry of the object. In the second image one of the four

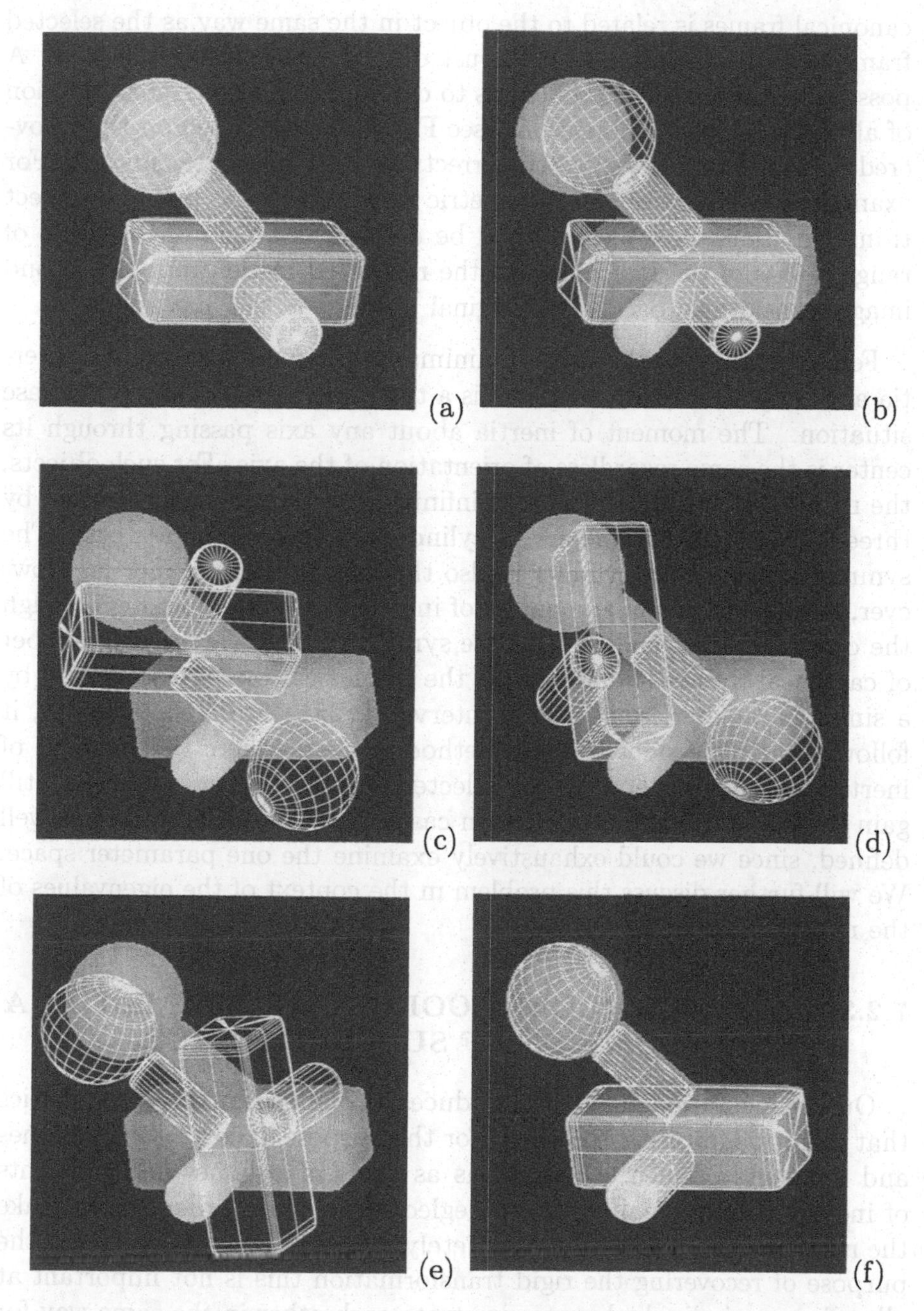

Figure 7.7. Registration based on inertial moments produces four possible solutions for the canonical frame: (a) *view-1* range image and recovered *model-1*, (b) *view-2* range image with overlaid *model-1* using unit transformation $\mathbf{T} = \mathbf{I}$, (c), (d), (e) and (f) represent *view-2* range image overlaid with *model-1* in four possible orientations that produce the same diagonal inertial matrix in the canonical frame. Only (f) turns out to be the right registration.

canonical frames is related to the object in the same way as the selected frame from the first image. We just do not know which one it is. A possible solution to this problem is to compute the rigid transformation of all four possible combinations (see Fig. 7.7), and based on the recovered transformations verify the correctness of each transformation. For example, we could use an error metric as a measure to find the correct transformation. The metric could be defined as a sum of distances of range points of the first image to the recovered model from the second image transformed back to its original position in the first image.

For some objects, the axes of minimal and maximal moment of inertia are not well defined. A sphere is a typical example of the worst case situation. The moment of inertia about any axis passing through its center is the same regardless of orientation of the axis. For such objects, the number of canonical frames is infinite, and can be parameterized by three independent parameters. A cylinder is a less degenerate case. The symmetry axis of the cylinder is also the axis of minimal inertia. However, the axis of maximal moment of inertia is any axis passing through the center and perpendicular to the symmetry axis. Again, the number of canonical frames is infinite, but the frames can be parameterized by a single parameter from a finite interval $[0, 2\pi)$. From the analysis, it follows that we cannot use the method in cases where the moment of inertia does not depend on the selected axis. However, we might still gain some additional information in cases where at least one axis is well defined, since we could exhaustively examine the one parameter space. We will further discuss this problem in the context of the eigenvalues of the matrix of inertia.

7.2.3.1 A CANONICAL COORDINATE FRAME FOR A COMPOSITION OF SUPERQUADRICS

Our segmentation algorithm produces a composition of superquadrics that might penetrate each other. For the purpose of computing volumes and moments of such compositions as sums of volumes and moments of inertia of individual parts, we neglect this fact in order not to make the method computationally completely infeasible. Nevertheless, for the purpose of recovering the rigid transformation this is not important at all as long as individual parts penetrate each other in the same way for each description which is recovered from range images. Of course, in practice, the recovered compositions are not likely to be equal in terms of superquadrics size, orientation, position, penetration, etc. But still we can assume that small changes in any of these parameters lead to small changes in the recovered canonical frame, if eigenvalues of the inertia matrix are sufficiently distinct. This gives the method the necessary robustness for practical use.

The volume, center of gravity, and the matrix of inertia of a composition of N_S superquadrics are computed as

$$\mathbf{i} = \sum_{i=1}^{N_S} \mathbf{T}_i \mathbf{i}_i^L , \tag{7.20}$$

and

$$\mathbf{I} = \sum_{i=1}^{N_S} map(\mathbf{T}_{tra,i}, \mathbf{T}_{rot\_sub,i} \mathbf{I}_i^L \mathbf{T}_{rot\_sub,i}^{-1}, \mathbf{T}_{rot,i} \mathbf{i}_i^L) , \tag{7.21}$$

where $\mathbf{T}_i$ is the homogeneous matrix that maps the coordinates from the local coordinate system of the i-th superquadric to the global coordinate frame of the range image scanner. We assume that the matrix $\mathbf{T}_i$ is decomposed into two matrices corresponding to pure rotation and pure translation as

$$\mathbf{T}_i = \mathbf{T}_{tra,i} \mathbf{T}_{rot,i} . \tag{7.22}$$

The $\mathbf{i}_i^L$ is a vector of the first moments and the volume of the superquadric in the local coordinate system, which originates in the center of gravity of the superquadric thus

$$\mathbf{i}_i^L = \begin{bmatrix} 0 \\ 0 \\ 0 \\ V_i \end{bmatrix} . \tag{7.23}$$

The volume V_i of i-th superquadric is computed from equation (2.59) in Chapter 2. Similarly the matrix $\mathbf{I}_i^L$ of moments of inertia of the i-th superquadric in the local coordinate frame is diagonal since the deviational moments equal 0 in this frame. The diagonal terms are computed from expressions in equations (2.67), (2.68), and (2.69) in Chaptor 2

$$\mathbf{I}_i^L = \begin{bmatrix} I_{xx,i}^L & 0 & 0 \\ 0 & I_{yy,i}^L & 0 \\ 0 & 0 & I_{zz,i}^L \end{bmatrix} . \tag{7.24}$$

Now, that we have easily computed the moments of the object from its superquadric description, we seek the canonical frame. To recover the translation part of the relationship between the global and the canonical frame of the object, we first construct a frame $C1$ in which the first order moments vanish

$$\mathbf{i}^{C1} = \begin{bmatrix} 0 \\ 0 \\ 0 \\ V \end{bmatrix} = \mathbf{T}_G^{C1}\mathbf{i} = \begin{bmatrix} 1 & 0 & 0 & p_x \\ 0 & 1 & 0 & p_y \\ 0 & 0 & 1 & p_z \\ 0 & 0 & 0 & 1 \end{bmatrix} \begin{bmatrix} I_x \\ I_y \\ I_z \\ V \end{bmatrix}. \qquad (7.25)$$

The solution for the transformation matrix that maps coordinates in the global coordinate system to coordinates in the $C1$ coordinate system is

$$\mathbf{T}_G^{C1} = \begin{bmatrix} 1 & 0 & 0 & -I_x/V \\ 0 & 1 & 0 & -I_y/V \\ 0 & 0 & 1 & -I_z/V \\ 0 & 0 & 0 & 1 \end{bmatrix}. \qquad (7.26)$$

The matrix of inertia in the $C1$ frame is equal to:

$$\mathbf{I}^{C1} = map(\mathbf{T}_G^{C1}, \mathbf{I}, \mathbf{i}). \qquad (7.27)$$

Now we look for a rotation that will rotate the $C1$ frame to frame C in such a way that the matrix $\mathbf{I}^C$ will become diagonal

$$\mathbf{I}^C = \begin{bmatrix} I_{xx}^C & 0 & 0 \\ 0 & I_{yy}^C & 0 \\ 0 & 0 & I_{zz}^C \end{bmatrix} = \mathbf{T}_{rot\_sub}\mathbf{I}^{C1}\mathbf{T}_{rot\_sub}^{-1}. \qquad (7.28)$$

The matrix $\mathbf{I}^C$ is a real symmetric matrix. The spectral theorem of linear algebra (Strang, 1988) states, that every real symmetric matrix $\mathbf{A}$ can be diagonalized by an orthogonal matrix $\mathbf{Q}$

$$\mathbf{D} = \mathbf{Q}^{-1}\mathbf{A}\mathbf{Q}. \qquad (7.29)$$

The columns of $\mathbf{Q}$ contain a complete set of orthonormal eigenvectors of $\mathbf{A}$. By comparing equations (7.28) and (7.29) we see that

$$\mathbf{T}_{rot\_sub} = \mathbf{Q}^{-1}. \qquad (7.30)$$

Thus, the solution to our problem always exists. The next problem is whether this solution is unique or not. The uniqueness of solution is related to uniqueness of the eigenvalues of the matrix $\mathbf{A}$.

If the eigenvalues are all distinct, the corresponding unit length eigenvectors are orthonormal and distinct up to a -1 scalar factor (Strang, 1988). This leads to $3! = 6$ possible permutations of the three selected eigenvectors, and $2^3 = 8$ possible matrices $\mathbf{Q}$ formed out of each selection. However, we choose to order eigenvectors in the ascending order of corresponding eigenvalues, which produces a unique permutation with eight possible matrices $\mathbf{Q}$ and aligns the x and z axis of the C along the axis of minimal and maximal moment of inertia, respectively. Out of these eight matrices, only four will have determinant equal to 1, the others will have determinant equal to -1, which implies an inversion operation that takes a right-handed coordinate system into a left-handed one. It is obvious that such a transformation does not correspond to a rigid transformation and thus solutions, where $\det \mathbf{Q} = -1$, are discarded.

If some or all of the eigenvalues are equal, there is an infinite number of eigenvectors corresponding to the non-unique eigenvalue. In case that the algebraic multiplicity of eigenvalue is 2, the eigenvectors span a plane, and we can choose any two orthonormal vectors lying in this plane as the eigenvectors. In case that the algebraic multiplicity is 3, any vector is an eigenvector, and we can choose any orthonormal base for the matrix $\mathbf{Q}$. In case of non-unique eigenvalues the number of matrices $\mathbf{Q}$ that diagonalized the matrix is infinite, and so is the number of canonical frames that we strive to construct. It is clear that we cannot use this method for objects whose principal axes of inertia are not well defined.

The canonical frame C is related to the global frame G, for a selected variant out of four possible matrices $\mathbf{T}_{rot}$ as

$$\mathbf{x}^C = \mathbf{T}_{rot}\mathbf{T}_G^{C1}\mathbf{x}^G = \mathbf{T}_G^C\mathbf{x}^G. \tag{7.31}$$

7.2.3.2 COMPUTING THE RIGID TRANSFORMATION FROM TWO CANONICAL FRAMES

Since a canonical frame is rigidly attached to the object, the coordinates of the same point expressed in any canonical frame are the same. Let us name the point as P and the canonical frame in the first image and the second image as $C1$ and $C2$, respectively (Fig. 7.8). So for the same object point P that has global coordinates $\mathbf{x}_1$ in the first frame, and global coordinates $\mathbf{x}_2$ in the second frame, the following relationships hold

$$\mathbf{x}_1 = \mathbf{T}_{C1}^G\mathbf{x}^{C1}, \tag{7.32}$$

$$\mathbf{x}_2 = \mathbf{T}_{C2}^G\mathbf{x}^{C2}, \tag{7.33}$$

$$\mathbf{x}^{C1} = \mathbf{x}^{C2}. \tag{7.34}$$

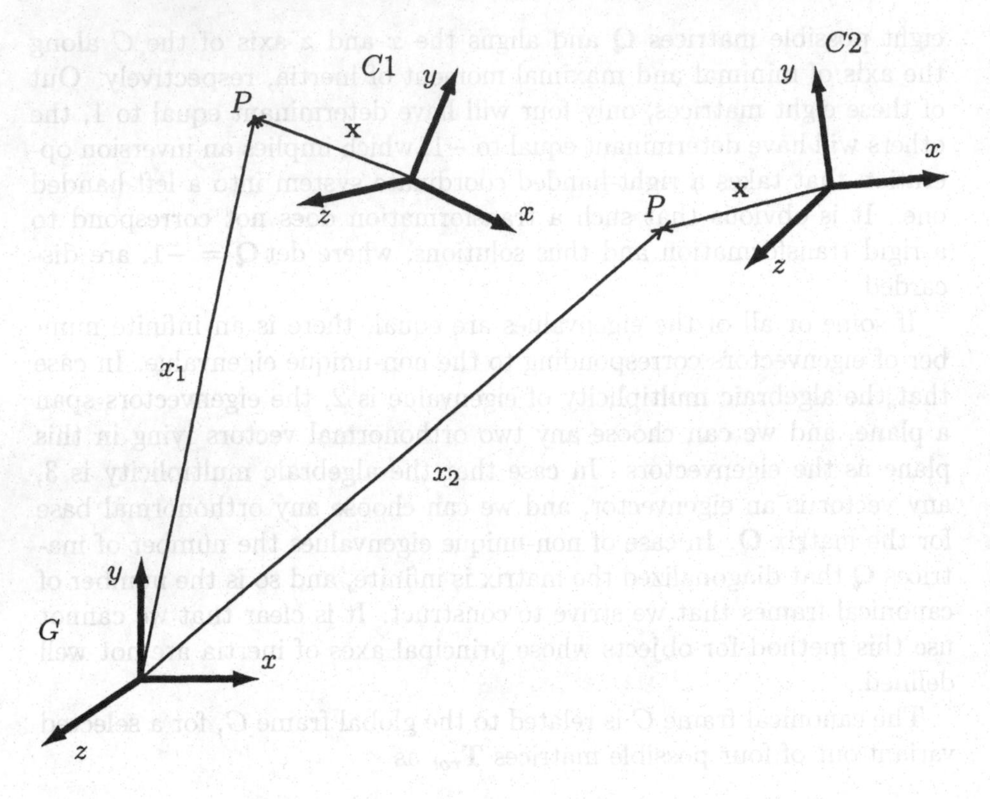

Figure 7.8. Vector **x** to a particular object point P, expressed in the canonical coordinate frame, is constant.

The rigid transformation of the point from the first to the second image is thus

$$\mathbf{x}_2 = \mathbf{T}\mathbf{x}_1 = \mathbf{T}_{C1}^{G}(\mathbf{T}_{C2}^{G})^{-1}\mathbf{x}_1, \tag{7.35}$$

so

$$\mathbf{T} = \mathbf{T}_{C1}^{G}(\mathbf{T}_{C2}^{G})^{-1}. \tag{7.36}$$

7.2.4 ESTIMATION OF RIGID TRANSFORMATIONS

To experimentally verify the proposed method, we generated a set of synthetic range images with known ground truth transformations (Fig. 7.9). We also compared the precision of estimates computed from moments of inertia of recovered models with the estimates based on moments of range image data points. In the latter case, we approximated the integrals needed in calculations of moments with sums of coordinates of a set of N range data points as follows

$$\mathbf{i} = \begin{bmatrix} \sum_{i=1}^{N} x_i \\ \sum_{i=1}^{N} y_i \\ \sum_{i=1}^{N} z_i \\ N \end{bmatrix}, \tag{7.37}$$

$$\mathbf{I} = \begin{bmatrix} \sum_{i=1}^{N}(y_i^2 + z_i^2) & -\sum_{i=1}^{N} x_i y_i & -\sum_{i=1}^{N} x_i z_i \\ -\sum_{i=1}^{N} x_i y_i & \sum_{i=1}^{N}(x_i^2 + z_i^2) & -\sum_{i=1}^{N} y_i z_i \\ -\sum_{i=1}^{N} x_i z_i & -\sum_{i=1}^{N} y_i z_i & \sum_{i=1}^{N}(x_i^2 + y_i^2) \end{bmatrix}. \tag{7.38}$$

We compared both methods qualitatively by overlaying a wireframe of recovered superquadric description from the first view over the range image from the second view, using the recovered transformation. For both methods, we selected the correct transformation out of the four possible transformations by hand.

Computing a residual rigid transformation that transforms an estimated transformation to the true transformation is one of the possible quantitative measures of precision of recovered transformation

$$\mathbf{T}_{true} = \mathbf{T}_{residual}\mathbf{T}_{estimate}. \tag{7.39}$$

Ideally, the residual transformation homogeneous matrix would equal a unit homogeneous matrix.

Figs. 7.10, 7.11, 7.12, 7.13 show the results of both methods graphically together with corresponding residual transformations, where for example, $T^{image}_{res\_v1\_v2}$ is a residual transformation for the method that transforms *view-1* to *view-2* based on range image points and $T^{model}_{res\_v1\_v2}$ is a residual transformation for the method that transforms *view-1* to *view-2* based on the recovered models.

Note the superior results of registration based on the moments of inertia of recovered superquadric models. The good performance of the method can be explained by its insensitivity to the density of range image data points, which depends on direction of local surface normal, and much lower sensitivity to self occlusion. The method could also be used for object pose estimation and recognition as described in (Galvez and Canton, 1993).

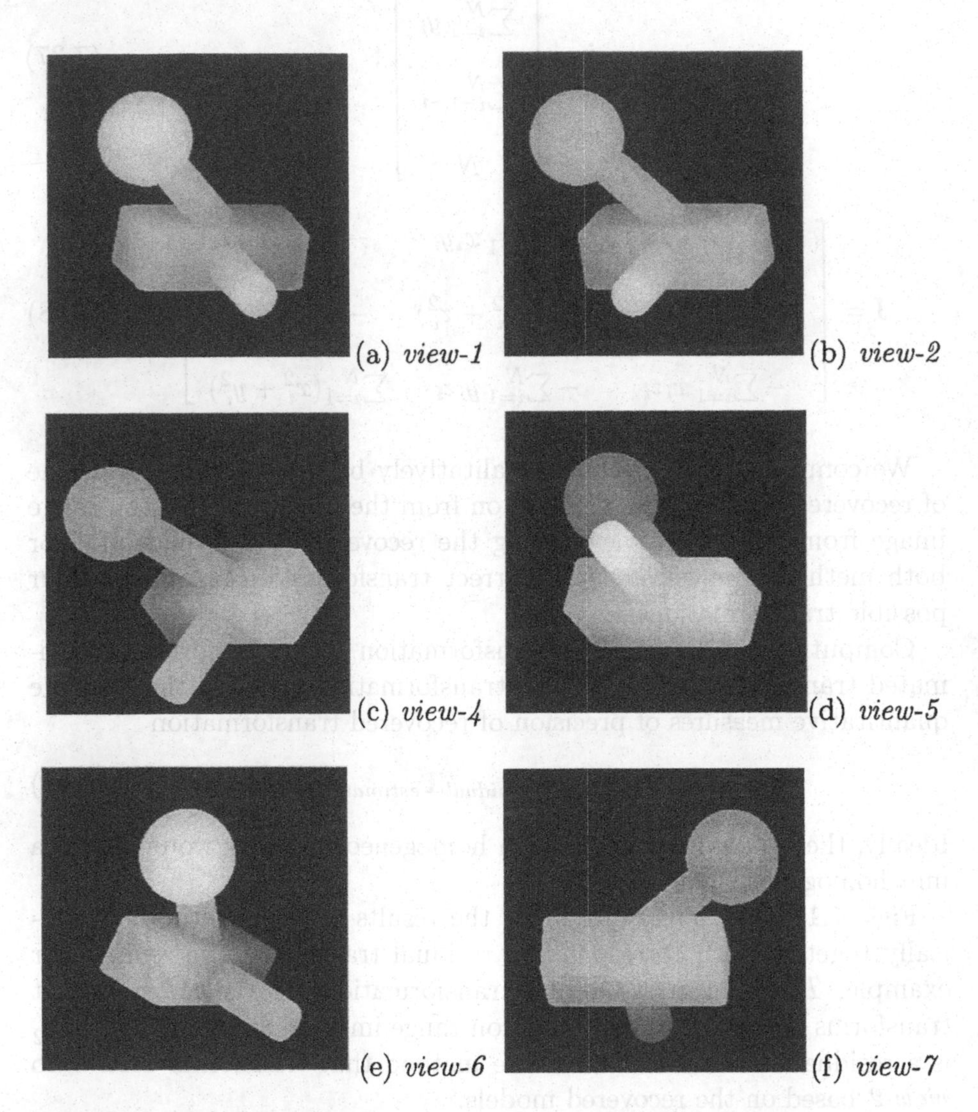

Figure 7.9. **A sequence of synthetic range images, with known rigid transformations, used in experimental comparison of the registration methods based on moments of inertia and center of gravity.**

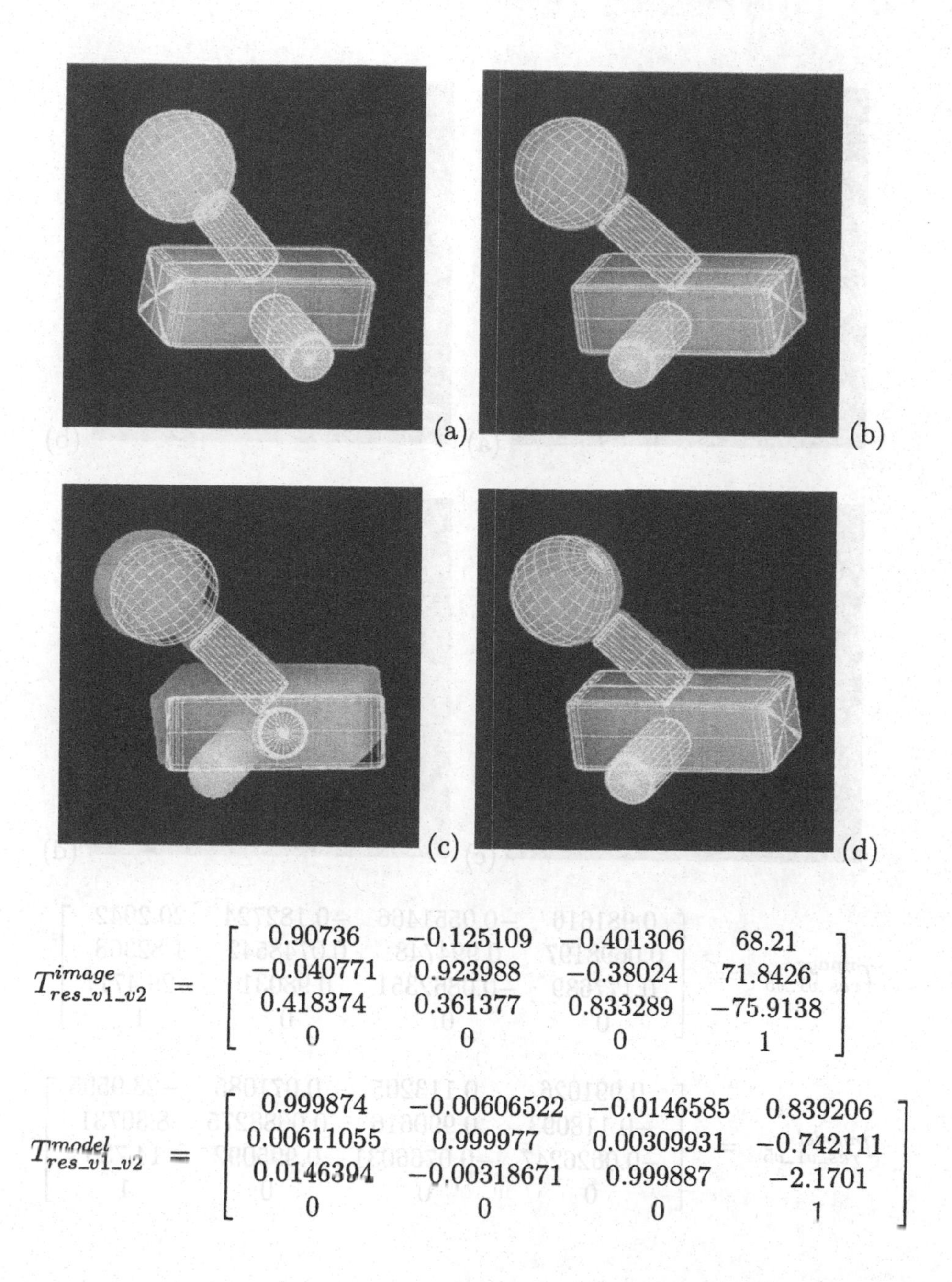

$$T^{image}_{res\_v1\_v2} = \begin{bmatrix} 0.90736 & -0.125109 & -0.401306 & 68.21 \\ -0.040771 & 0.923988 & -0.38024 & 71.8426 \\ 0.418374 & 0.361377 & 0.833289 & -75.9138 \\ 0 & 0 & 0 & 1 \end{bmatrix}$$

$$T^{model}_{res\_v1\_v2} = \begin{bmatrix} 0.999874 & -0.00606522 & -0.0146585 & 0.839206 \\ 0.00611055 & 0.999977 & 0.00309931 & -0.742111 \\ 0.0146394 & -0.00318671 & 0.999887 & -2.1701 \\ 0 & 0 & 0 & 1 \end{bmatrix}$$

Figure 7.10. Registration of range images *view-1* and *view-2*: (a) recovered *model-1* from *view-1*, (b) recovered *model-2* from *view-2*, (c) registration of *model-1* to *view-2* based on raw image data, (d) registration of *model-1* to *view-2* based on recovered superquadric models.

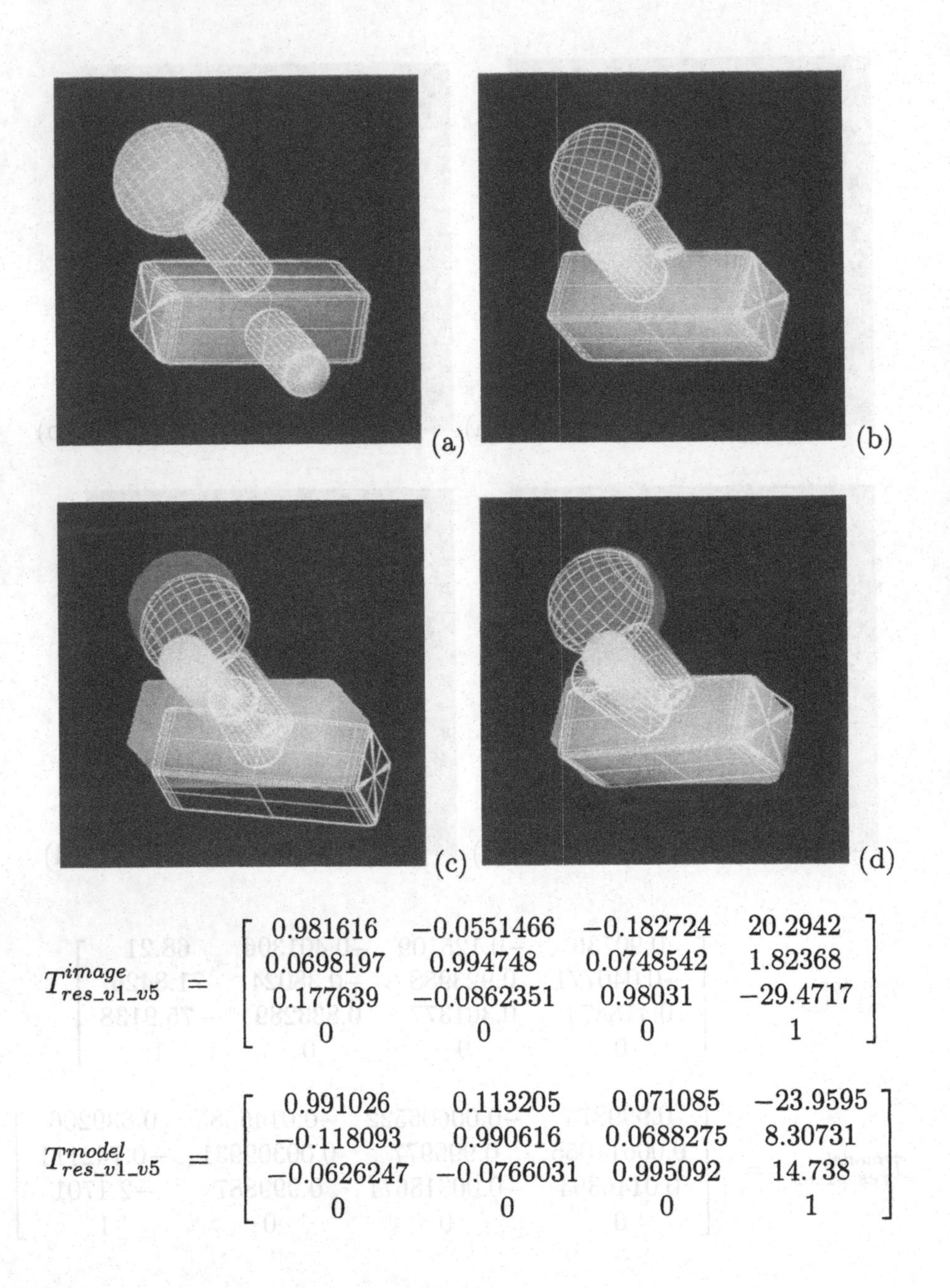

$$T^{image}_{res\_v1\_v5} = \begin{bmatrix} 0.981616 & -0.0551466 & -0.182724 & 20.2942 \\ 0.0698197 & 0.994748 & 0.0748542 & 1.82368 \\ 0.177639 & -0.0862351 & 0.98031 & -29.4717 \\ 0 & 0 & 0 & 1 \end{bmatrix}$$

$$T^{model}_{res\_v1\_v5} = \begin{bmatrix} 0.991026 & 0.113205 & 0.071085 & -23.9595 \\ -0.118093 & 0.990616 & 0.0688275 & 8.30731 \\ -0.0626247 & -0.0766031 & 0.995092 & 14.738 \\ 0 & 0 & 0 & 1 \end{bmatrix}$$

Figure 7.11. Registration of range images *view-1* and *view-5*: (a) recovered *model-1* from *view-1*, (b) recovered *model-5* from *view-5*, (c) registration of *model-1* to *view-5* based on raw image data, (d) registration of *model-1* to *view-5* based on recovered superquadric models.

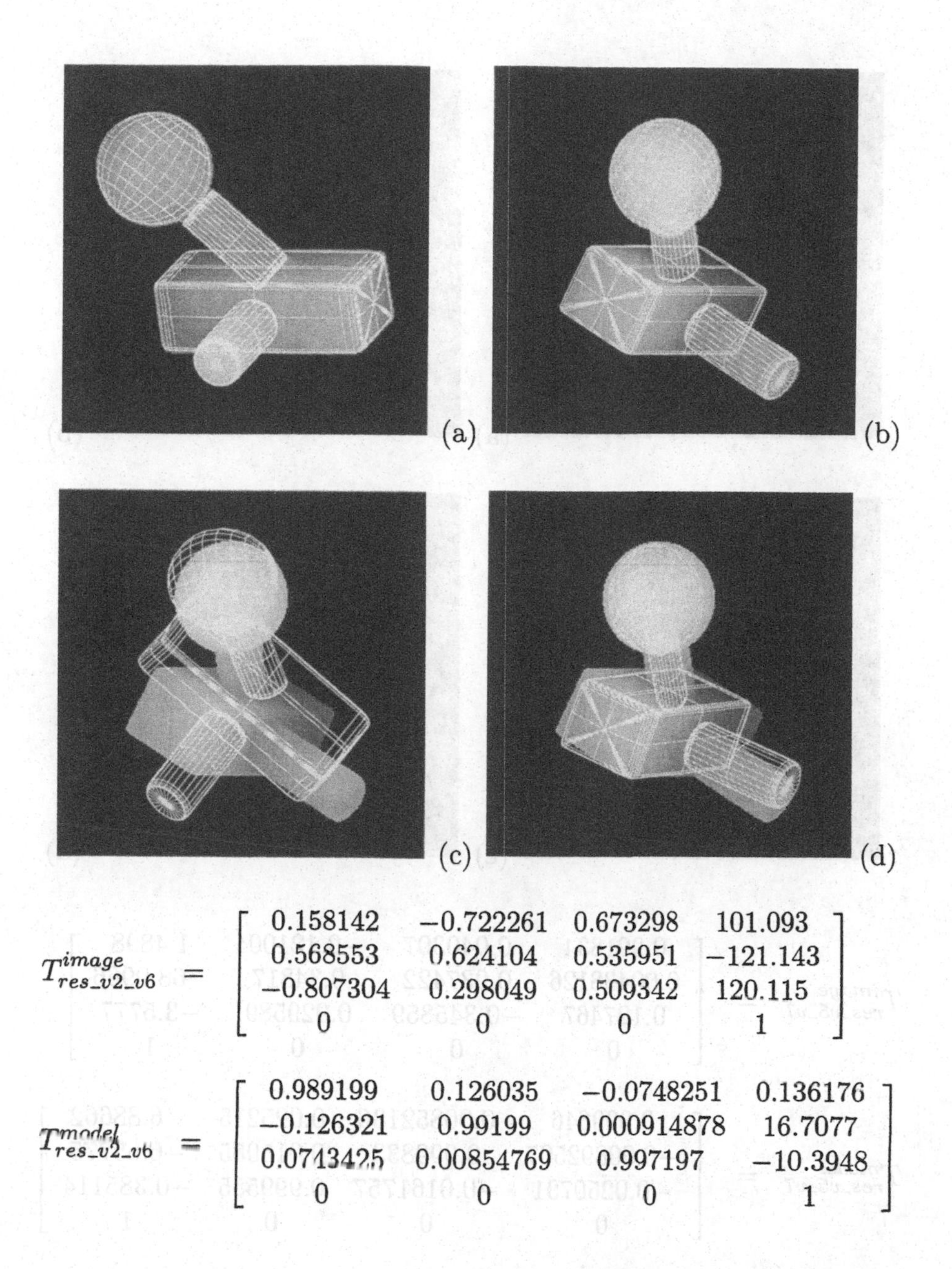

$$T^{image}_{res\_v2\_v6} = \begin{bmatrix} 0.158142 & -0.722261 & 0.673298 & 101.093 \\ 0.568553 & 0.624104 & 0.535951 & -121.143 \\ -0.807304 & 0.298049 & 0.509342 & 120.115 \\ 0 & 0 & 0 & 1 \end{bmatrix}$$

$$T^{model}_{res\_v2\_v6} = \begin{bmatrix} 0.989199 & 0.126035 & -0.0748251 & 0.136176 \\ -0.126321 & 0.99199 & 0.000914878 & 16.7077 \\ 0.0743425 & 0.00854769 & 0.997197 & -10.3948 \\ 0 & 0 & 0 & 1 \end{bmatrix}$$

Figure 7.12. Registration of range images *view-2* and *view-6*: (a) recovered *model-2* from *view-2*, (b) recovered *model-6* from *view-6*, (c) registration of *model-2* to *view-6* based on raw image data, (d) registration of *model-2* to *view-6* based on recovered superquadric models.

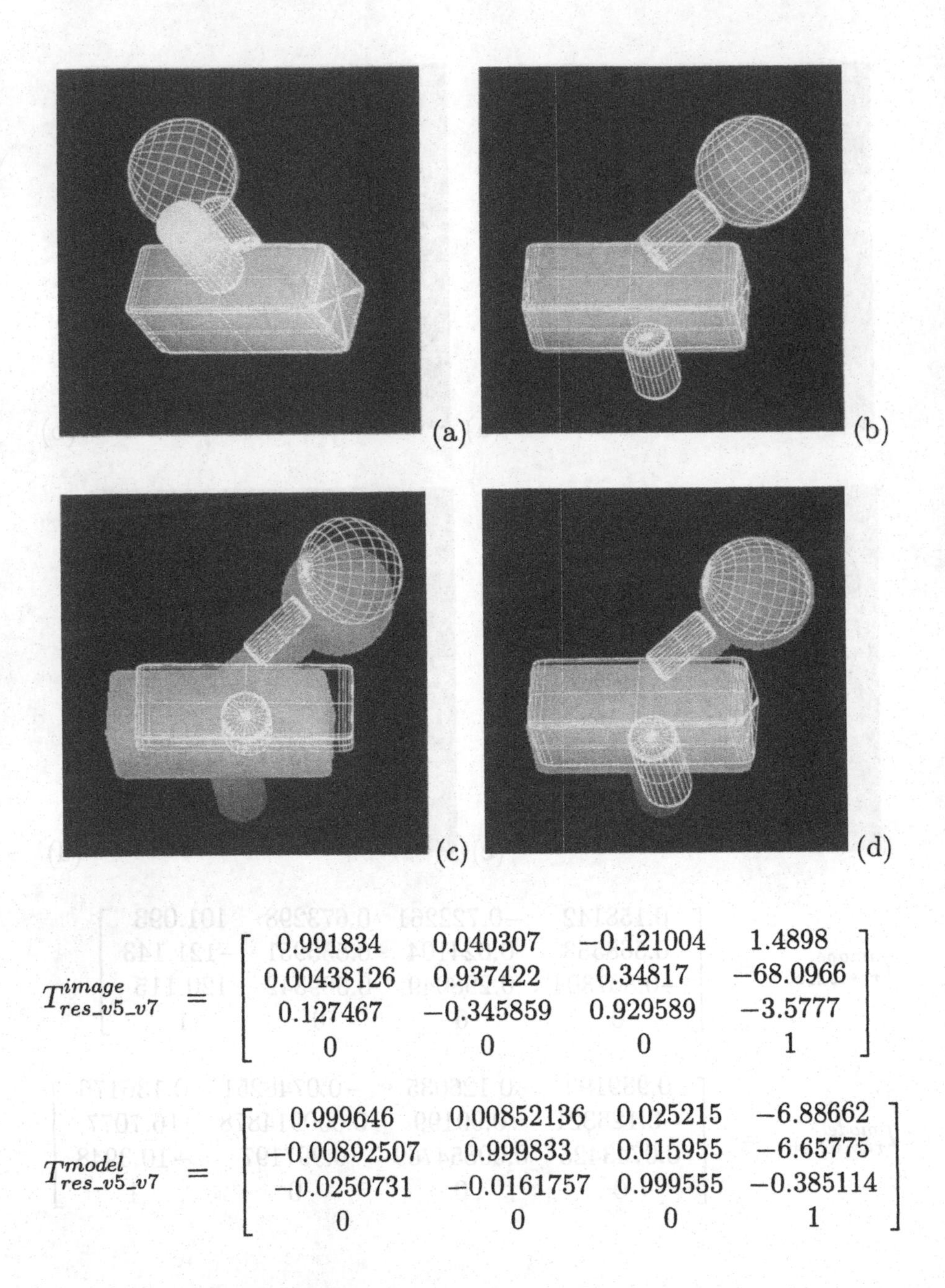

$$T_{res\_v5\_v7}^{image} = \begin{bmatrix} 0.991834 & 0.040307 & -0.121004 & 1.4898 \\ 0.00438126 & 0.937422 & 0.34817 & -68.0966 \\ 0.127467 & -0.345859 & 0.929589 & -3.5777 \\ 0 & 0 & 0 & 1 \end{bmatrix}$$

$$T_{res\_v5\_v7}^{model} = \begin{bmatrix} 0.999646 & 0.00852136 & 0.025215 & -6.88662 \\ -0.00892507 & 0.999833 & 0.015955 & -6.65775 \\ -0.0250731 & -0.0161757 & 0.999555 & -0.385114 \\ 0 & 0 & 0 & 1 \end{bmatrix}$$

Figure 7.13. Registration of range images *view-5* and *view-7*: (a) recovered *model-5* from *view-5*, (b) recovered *model-7* from *view-7*, (c) registration of *model-5* to *view-7* based on raw image data, (d) registration of *model-5* to *view-7* based on recovered superquadric models.

7.3 SUMMARY

Superquadrics can be used for a large variety of applications. In the first part of the chapter, there is an overview of applications using superquadric models. Superquadric models can be applied as generic volumetric shape primitives for shape representation, segmentation, object categorization, and object recognition. Application areas range from engineering to medicine.

Superquadrics are finding a role in robotics as part of dedicated vision systems, or for modeling the shape of robot kinematic chains and the robot workspace. The advantage of path planning with superquadrics is fast and simple collision detection, based on the implicit superquadric function which can be extended to a potential field. Using the same geometric models can enable a simpler fusion of visually derived information with the information of the robot control systems. Due to their modeling flexibility which can combine shape and physical properties, superquadrics are also a good vehicle for reasoning about function.

In the second part of the chapter, we show how superquadrics can be used for range image registration. The closed form expressions for computation of moments of inertia for superquadrics in their canonical coordinate system, and expressions for transformation of the moments to other coordinate systems provide an efficient way of computing moments of inertia of objects composed of several superquadrics. Similarly, the expressions for volume of a superquadric leads to simple expressions for the center of gravity of such compositions. Both results, the center of gravity and matrix of moments of inertia, can be used to compute a rigid transformation between two range images, which have been modeled by superquadrics. We demonstrate our method of registration on a set of synthetic range images with known ground truth. Our method using superquadric models performs significantly better than the same registration method that uses only raw range point data.

Chapter 8

CONCLUSIONS

Recovery of part-level shape models from images is very important for object recognition, manipulation, object avoidance, part-based segmentation, and other tasks which rely on a compact and generic shape description. Among several part-level models that have been proposed in computer vision, generalized cylinders and superquadrics are the two most widely used models.

In computer vision a particular shape model is chosen according to its degree of uniqueness and compact representation, its local support, expressiveness, and preservation of information. Generalized cylinders are somewhat hindered by the difficulty of their recovery from images. Superquadrics, on the other hand, are not as powerful in representing different shapes, but have some other special advantages:

- a small number of model parameters with intuitive meaning which have a large expressive power for natural shapes with rounded edges and corners, as well as for standard geometric solids with sharp edges,
- closed form expressions for geometric properties such as volume and moments of inertia,
- there exist robust methods for reconstruction from range images, and
- superquadrics can be enhanced by adding global and local shape deformations.

Superquadrics are an extension of quadric surfaces that can model a large variety of shapes which are useful for volumetric part representation of natural and man-made objects. They are defined by either one of two equivalent equations: the implicit form of the superquadric equation is important for the recovery of superquadrics and testing for intersection, while the explicit form is more suitable for rendering. Geometric properties such as radial Euclidean distance from a point to the superellipsoid surface, superellipsoid volume and superellipsoid moments of inertia, which we derived, are used for recovery of superquadrics and for range image registration.

The expressive power of superquadrics can be increased with global and local shape deformations. With the aid of deformations, superquadrics can better and more closely model various natural and man-made shapes. Global deformations enhance the generic part-level modeling capability of superquadrics. Global deformations which act on the whole model, with just a few global deformation parameters, can be tightly integrated with superquadric parameterization and recovery. We defined global tapering and bending of superquadrics which in combination require just four additional parameters. Local deformations, on the other hand, wrap around the superquadric an additional parametric layer to enable fine-grained local changes of shape. By their nature, the definition of local deformations requires a larger number of parameters. Superquadrics with local deformations have usually more degrees of freedom than there are data points, so that the recovery process is underconstrained and therefore the recovered models are not unique.

The most decisive test of the actual usefulness of a chosen shape model in computer vision is, how easy it is to recover the model from images. We describe in the book several methods of superquadric recovery that differ on:

input data, which can take the form of 3D points (range data, stereo, tactile information), 2D contours (silhouettes, edges) and of surface normal information (shape from shading),

solution types, ranging from analytical solutions, minimization of objective functions, matched filters, interval bi-section methods, point distribution models, and physics-based methods which address in particular the recovery of locally deformed superquadrics,

objective functions, which use ad-hoc definitions such as counting the number of overlapping pixels or those which are based on algebraic distance, radial Euclidean distance, and surface normal information, and

minimization methods. The minimization of non-linear cost functions can be solved by search, numerical methods, different analytical gradient-descent methods (Levenberg-Marquardt or Gauss-Newton), simulated annealing, and genetic algorithms.minimization,simulated annealing

In particular, we give a detailed description of our own least-squares gradient descent minimization technique for recovery of superquadrics from pre-segmented range images. We also show how the recovery method can also be extended to recovery of globally deformed superquadrics. For minimization, we use a cost function which is based on the inside-outside function, and additionally weighted by the volume of the superquadric.

Most superquadric recovery methods assume pre-segmented range data. We show how segmentation and recovery of parametric models can be combined. The major novelty of our *recover-and-select* segmentation scheme is that the parametric models are simultaneously and independently recovered by growing models from the initial *seeds*, which results in a robust, but redundant representation. From this redundant representation we obtain the final description by solving an optimization problem, cast as a search for the compact MDL-like representation. Optimization of the objective function that encompasses information about the competing models is performed by a simple winner-takes-all algorithm which turns out to be a good compromise between speed and precision. The computational complexity of the overall segmentation and recovery method can be reduced by interleaving the model-recovery and model-selection procedures. Since it is computationally prohibitive to grow all possible hypothesized part-models to their full extent, and then select the final models, best models are selected after every few iterations of model recovery. Superquadric model extrapolation (growing) is performed by dilating the superquadric model in all directions, while new compatible data points are accepted on the basis of the radial Euclidean distance.

Experimental results of segmentation and recovery of superquadrics from range data show that we have successfully combined the two existing methods, namely recovery of superquadric models (Solina, 1987) and the *recover-and-select* segmentation paradigm (Leonardis, 1993). We demonstrate that a direct segmentation of range images into part-level volumetric models, which is stable with regard to the number of parts and their generic shape, is possible. The accuracy of individual superquadric models can be increased by a post-processing method. The recover-and-select segmentation with superquadrics is also stable with regard to changing viewpoint. At the same time, it successfully eliminates outliers which are present in range images. The method degrades gracefully if the data can not be modeled by superquadrics. In such cases, enriching the shape vocabulary helps, either by introducing global and local deformations of superquadrics or by combining different types of models, such as surface patches and superquadrics, for example.

Superquadrics can be used for a large variety of applications: as generic volumetric shape primitives for shape representation, segmentation, object categorization, and object recognition. Application areas range from engineering to medicine. Superquadrics are used in robotics for modeling the shape of robot kinematic chains and the robot workspace. Path planning with superquadrics is fast and simple collision detection can be implemented with the implicit superquadric function which can be extended to a potential field.

We developed a range image registration method based on superquadric representation of range images (Jaklič, 1997). We compute the moments of inertia and volume of objects composed of several superquadrics. The center of gravity and matrix of moments of inertia are used to compute a rigid transformation between two range images, which have been modeled by superquadrics. The testing of this range image registration method the method performs much better than the registration that uses only raw range point data sets.

Superquadrics are becoming a fairly standard model for range image interpretation in computer vision since the recovery and segmentation of globally deformed superquadrics from range images is relatively well understood. For the recovery of superquadrics, one wishes a better error-of-fit cost function which would weight uniformly all data points, which would possibly take into account also the distribution of measurement errors, and at the same time make fast and reliable minimization possible. Recovery and segmentation of superquadrics from intensity images, on the other hand, is much more difficult. To increase the expressive power, superquadrics are enhanced with global and local deformations. What is the best hierarchical parameterization of global and local deformations which would enable several levels of shape abstraction, and at the same time, a reliable recovery from image data, is still an open problem. Unresolved is also the generation of canonical superquadric shape descriptions which are needed for database indexing and object recognition. Nonetheless, advances in superquadric research are fast and we hope that most open problems will be resolved in the near future.

Appendix A
Rendering of Superquadrics in *Mathematica*

Rendering of superquadrics using the explicit function of the superquadric surface is straightforward. In the *Mathematica* environment one can simply use the built-in function

$$\texttt{ParametricPlot3D}[\{f_x, f_y, f_z\}, \{t, t_{min}, t_{max}\}, \{u, u_{min}, u_{max}\}].$$

ParametricPlot3D draws the surface as a function of two parameters t in u. Using the explicit (parametric) equation of the superquadric (see Eq. (2.10) on page 19), we write

```
SQParametricPlot[{e1_,e2_,a1_,a2_,a3_}] :=
  ParametricPlot3D[
  {
    a1 Sign[Cos[eta] Cos[omega]] Abs[Cos[eta]]^e1 Abs[Cos[omega]]^e2,
    a2 Sign[Cos[eta] Sin[omega]] Abs[Cos[eta]]^e1 Abs[Sin[omega]]^e2,
    a3 Sign[Sin[eta]] Abs[Sin[eta]]^e1
  },
  {eta,- Pi/2,Pi/2},
  {omega,- Pi,Pi}]
```

The *Mathematica* interface enables the user to view the superquadric from any direction.

To draw a superquadric in general position, we need three additional parameters to define its position in a global coordinate system p_x, p_y, p_z and three to define its orientation r_x, r_y, r_z (i.e., the Euler angles ϕ, θ, ψ). Using homogeneous coordinates, we multiply the position vector $\mathbf{x}(\eta, \omega)$ given in equation (2.10) with the homogeneous transformation matrix **T**, (Eq. 2.34)

$$\mathbf{T} = \begin{bmatrix} \cos\phi\cos\theta\cos\psi - \sin\phi\sin\psi & -\cos\phi\cos\theta\sin\psi - \sin\phi\cos\psi & \cos\phi\sin\theta & p_x \\ \sin\phi\cos\theta\cos\psi + \cos\phi\sin\theta & -\sin\phi\cos\theta\sin\psi + \cos\phi\cos\theta & \sin\phi\sin\theta & p_y \\ -\sin\theta\cos\psi & \sin\theta\sin\psi & \cos\theta & p_z \\ 0 & 0 & 0 & 1 \end{bmatrix}.$$

The function for drawing a superquadric in general position in *Mathematica* is therefore:

```
SQParametricPlot[{e1_,e2_,a1_,a2_,a3_,rx_,ry_,rz_,px_,py_,pz_}] :=
    ParametricPlot3D[ {
                        a1 Sign[Cos[eta] Cos[omega]] Abs[Cos[eta]]^e1 \
                        Abs[Cos[omega]]^e2 \
                        (Cos[rx] Cos[ry] Cos[rz] - Sin[rx] Sin[rz]) \
                        + a2 Sign[Cos[eta] Sin[omega]] Abs[Cos[eta]]^e1 \
                        Abs[Sin[omega]]^e2 \
                        (-  Cos[rx] Cos[ry] Sin[rz] - Sin[rx] Cos[rz]) \
                        + a3 Sign[Sin[eta]] Abs[Sin[eta]]^e1 \
                        Cos[rx] Sin[ry] + px,

                        a1 Sign[Cos[eta] Cos[omega]] Abs[Cos[eta]]^e1 \
                        Abs[Cos[omega]]^e2 \
                        Sin[rx] Cos[ry] Cos[rz] + Cos[rx] Sin[rz] \
                        + a2 Sign[Cos[eta] Sin[omega]] Abs[Cos[eta]]^e1 \
                        Abs[Sin[omega]]^e2 \
                        (-  Sin[rx] Cos[ry] Sin[rz] + Cos[rx] Cos[rz]) \
                        + a3 Sign[Sin[eta]] Abs[Sin[eta]]^e1 \
                        Sin[rx] Sin[ry] + py,

                        a1 Sign[Cos[eta] Cos[omega]] Abs[Cos[eta]]^e1 \
                        Abs[Cos[omega]]^e2 \
                         -  Sin[ry] Cos[rz] \
                        + a2 Sign[Cos[eta] Sin[omega]] Abs[Cos[eta]]^e1 \
                        Abs[Sin[omega]]^e2 \
                        Sin[ry] Sin[rz] \
                        + a3 Sign[Sin[eta]] Abs[Sin[eta]]^e1 \
                        Cos[ry] + pz

                    },
                    {eta,- Pi/2,Pi/2},

                    {omega,- Pi,Pi}];
```

Appendix B
Superquadric Recovery Code

Software for least-squares based superquadric recovery from pre-segmented range images which is described in Chapter 4 of this book and was developed at the University of Pennsylvania, can be obtained from `ftp.cis.upenn.edu` via anonymous `ftp`. The superquadric library is in `/pub/grasp/sq.tar.Z`. The code is written in C. The library can be compiled with the PM image format or without it. PM is the image header format used in the GRASP laboratory of the University of Pennsylvania. Using PM format allows the range images of the objects to be directly input to the superquadric recovery program. If PM format is not used, then a list of points in the world coordinate system is required by the program. The PM library can be copied from the `/pub/grasp/pm2` directory.

This code has been modified and integrated in the *Segmentor* software package (see Appendix E).

Appendix B
Superquadric Recovery Code

Software for least-squares based superquadric recovery from pre-segmented range images which is described in Chapter 4 of this book and was developed at the University of Pennsylvania, can be obtained from ftp.cis.upenn.edu via anonymous ftp. The superquadric library is in /pub/grasp/sq.tar.Z. The code is written in C. The library can be compiled with the PM image format or without it. PM is the image header format used in the GRASP laboratory of the University of Pennsylvania. Using PM format allows the range images of the objects to be directly input to the superquadric recovery program. If PM format is not used, then a list of points in the world coordinate system is required by the program. The PM library can be copied from the /pub/grasp/pm2 directory.

This code has been modified and integrated in the *Segmentor* software package (see Appendix E).

Appendix C
Range Image Acquisition

We acquired most of the range images for the experiments in this book with a structured light range scanner (Skočaj, 1999). Fig. C.1 shows the system configuration. We use a coded light range sensor based on an LCD stripe projector (Fig. C.2). The projector projects nine Gray coded stripe patterns onto the object which is placed into the work space of the range scanner.

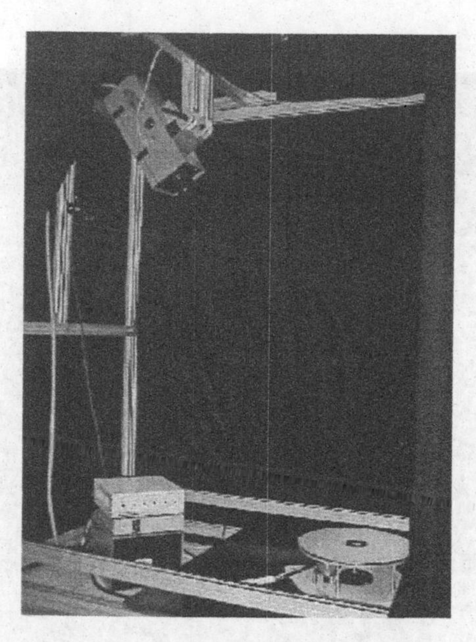

Figure C.1. The range scanner system showing the camera and the light stripe projector in the upper left corner of the system and the workspace with a turn-table in the lower right corner.

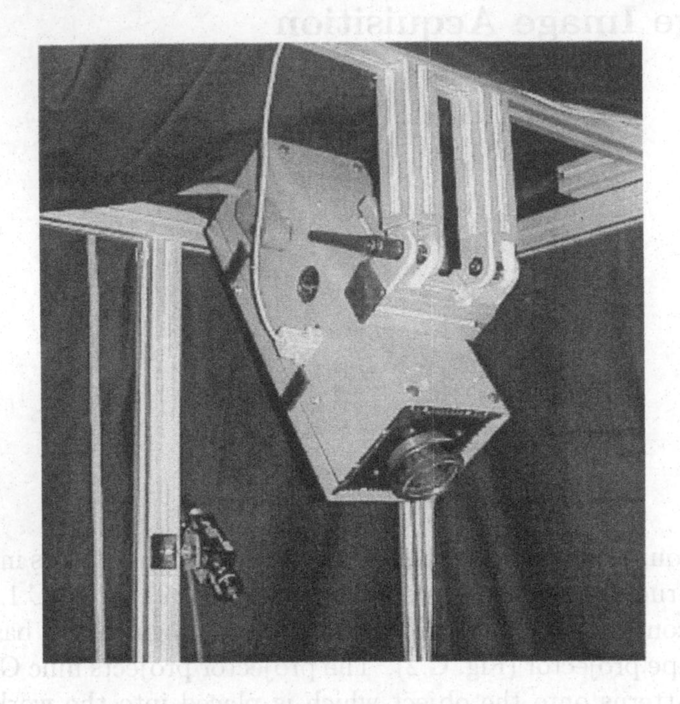

Figure C.2. A closer view of the camera and the light stripe projector of the range scanner system.

Figure C.3. One of several stripe patterns projected onto the object.

Fig. C.3 shows one of the stripe patterns which is projected onto the object. With the stripe patterns, we can distinguish 2^9 different projection planes indicated by s coordinate in the projector space (Fig. C.4). A CCD camera which is mounted below the projector acquires grey level images of the deformed stripe patterns, which are binarized so as to separate projector-illuminated from non-illuminated areas. For each pixel, a 9-bit code is stored which represents the s coordinate of the point in the projector space. If the calibration parameters of the range scanner setup are known, we can compute the 3-D coordinates of a point in the range scanner work space from the s coordinate in the projector space and the u and v coordinates of the corresponding pixel in the image space by using active triangulation.

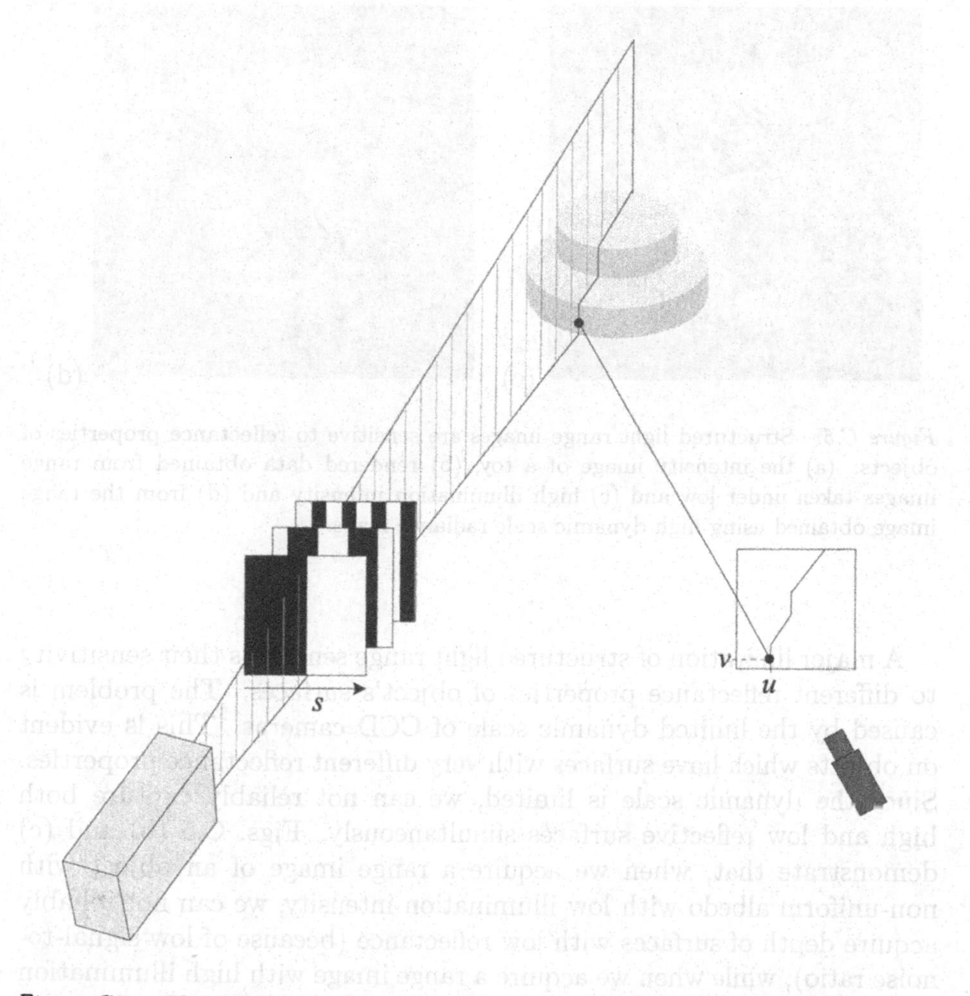

Figure C.4. **The acquisition principle of the structured light range scanner.**

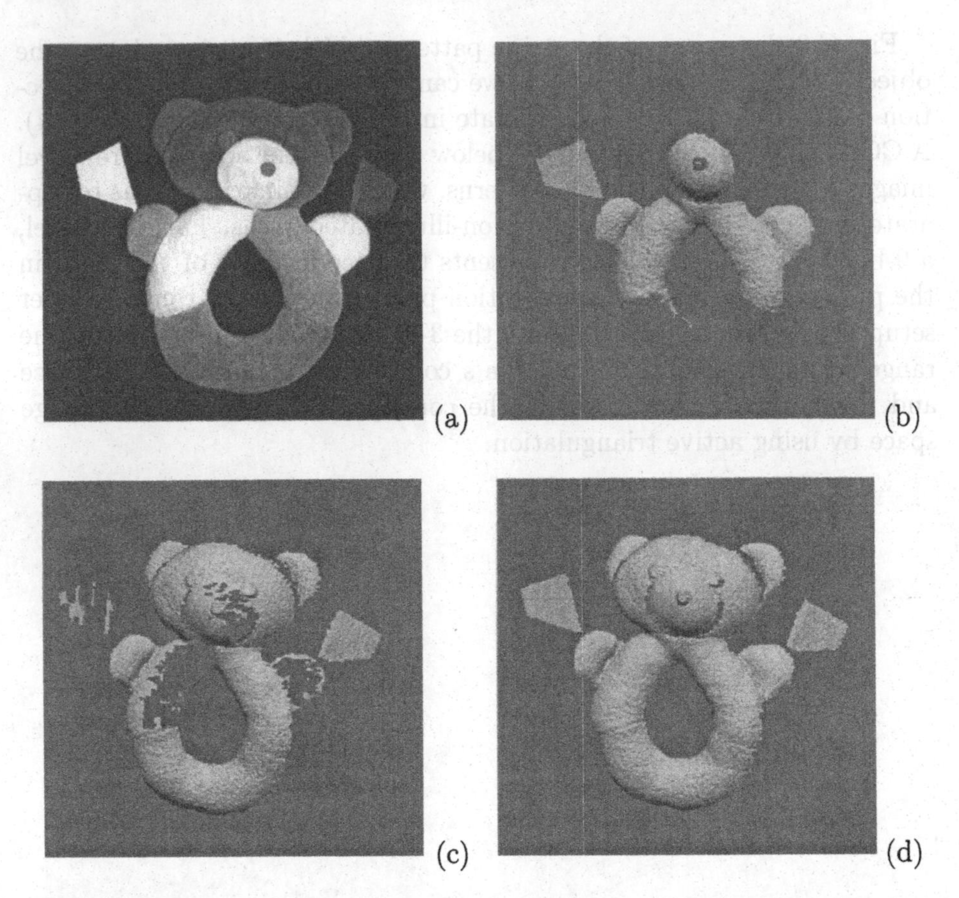

Figure C.5. Structured light range images are sensitive to reflectance properties of objects: (a) the intensity image of a toy, (b) rendered data obtained from range images taken under low and (c) high illumination intensity and (d) from the range image obtained using high dynamic scale radiance maps.

A major limitation of structured light range sensors is their sensitivity to different reflectance properties of object's surfaces. The problem is caused by the limited dynamic scale of CCD cameras. This is evident on objects which have surfaces with very different reflectance properties. Since the dynamic scale is limited, we can not reliably capture both high and low reflective surfaces simultaneously. Figs. C.5 (b) and (c) demonstrate that, when we acquire a range image of an object with non-uniform albedo with low illumination intensity, we can not reliably acquire depth of surfaces with low reflectance (because of low signal-to-noise ratio), while when we acquire a range image with high illumination intensity, we can not recover depth of surfaces with high reflectance (because of the sensor saturation).

To overcome this problem, we take several intensity images under different illumination intensities and combine them into a high dynamic scale radiance map (Skočaj and Leonardis, 2000). High dynamic scale radiance maps are then used instead of original intensity images as an input to the algorithm for range image formation. In such a way, we overcome the problem caused by the limited scale of intensity images, and we can reliably capture range images of objects with non-uniform albedo (Fig. C.5 (d)).

Appendix D
Minimum Description Length and Maximum A Posteriori Probability

We briefly review the equivalence between the Minimum Description Length (MDL) principle and the Maximum A Posteriori (MAP) probability principle. For a more elaborate discussion on this topic, the interested reader is referred to (Rissanen, 1983; Rissanen, 1984; Leclerc, 1989a; Keeler, 1991).

Consider a source on an alphabet X of observations x whose distribution is defined by a parameter θ in an alphabet Θ. Let $p_\Theta(\theta)$ be a prior distribution of the parameter θ, and let the conditional probability be given as $p_{X|\Theta}(x|\theta)$. According to Shannon, there is a correspondence between the probability distribution and the length of encoding. If the model and the data are described by discrete probabilities, then the total length $L(\cdot)$ of the codeword for (x, θ) is

$$L_{X,\Theta} = L_{X|\Theta}(x|\theta) + L_\Theta(\theta) = -log_2 p_{X|\Theta}(x|\theta) - log_2 p_\Theta(\theta) \ . \qquad \text{(D.1)}$$

According to the MDL principle, the task is to find the parameter θ which minimizes the expected length of the encoding, i.e., the *MDL estimator*

$$\hat{\theta} = \min_{\theta \in \Theta} L_{X,\Theta} = \min_{\theta \in \Theta}(L_{X|\Theta}(x|\theta) + L_\Theta(\theta)) \ . \qquad \text{(D.2)}$$

Due to the monotonicity of the logarithmic function, we can write the equation (D.2) as

$$\hat{\theta} = \min_{\theta \in \Theta} L_{X,\Theta} = \max_{\theta \in \Theta}(p_{X|\Theta}(x|\theta) p_\Theta(\theta)) \ . \qquad \text{(D.3)}$$

Let us now consider a more traditional MAP estimation, which is a direct application of Bayes' rule,

$$p_{(\Theta|X)}(\theta|x) = \frac{p_{X|\Theta}(x|\theta)p_\Theta(\theta)}{p_X(x)} \ . \tag{D.4}$$

We choose the value of the parameter $\theta \in \Theta$ which maximizes the *a posteriori* probability of the parameter θ

$$\hat{\theta} = \max_{\theta\in\Theta} p_{\Theta|X}(\theta|x) = \max_{\theta\in\Theta} \frac{p_{X|\Theta}(x|\theta)p_\Theta(\theta)}{p_X(x)} \ . \tag{D.5}$$

Since $p_X(x) = \sum_{\theta\in\Theta} p_{X|\Theta}(x|\theta)p_\Theta(\theta)$ is constant, maximizing equation (D.5) is equivalent to maximizing

$$\hat{\theta} = \max_{\theta\in\Theta} p_{\Theta|X}(\theta|x) = \max_{\theta\in\Theta}(p_{X|\Theta}(x|\theta)p_\Theta(\theta)) \ , \tag{D.6}$$

which is identical to equation (D.3).

By choosing the optimal (shortest) descriptive language for a given probability distribution, the MDL criterion is equivalent to the maximizing of the MAP. The equivalence holds in both directions, namely, if the probability distribution is implicitly defined on a given descriptive language, the MAP principle is equivalent to the search for the MDL.

Appendix E
Object-Oriented Framework for Segmentation (Segmentor)

Segmentor is an object-oriented framework for image segmentation based on the recover-and-select paradigm. It uses data abstraction, inheritance, and polymorphism to simplify and speed up application of the paradigm to different image domains using various image models.

Adaptation of the paradigm requires experimentation in finding the appropriate key elements of the paradigm, namely: parametric image models, distance functions between models and image values, model parameter estimation algorithms, algorithms for growing model descriptions, functions measuring the quality of descriptions and algorithms for selection of descriptions.

Since the recover-and-select paradigm is an iterative procedure, visualization of iterative steps during the segmentation process provides the experimenter with the critical feedback needed to asses the quality and suitability of chosen elements of the paradigm. Visualisation is also of utmost importance in testing and debugging experimental elements of the paradigm, as well as the framework itself.

We selected C++ as an object oriented programming language for the implementation and X Window System to implement the user interface. The former decision allows us to encapsulate and reuse existing software for numerical optimization written in C. The source code for *Segmentor* can be obtained from: `http://lrv.fri.uni-lj.si/~alesj`.

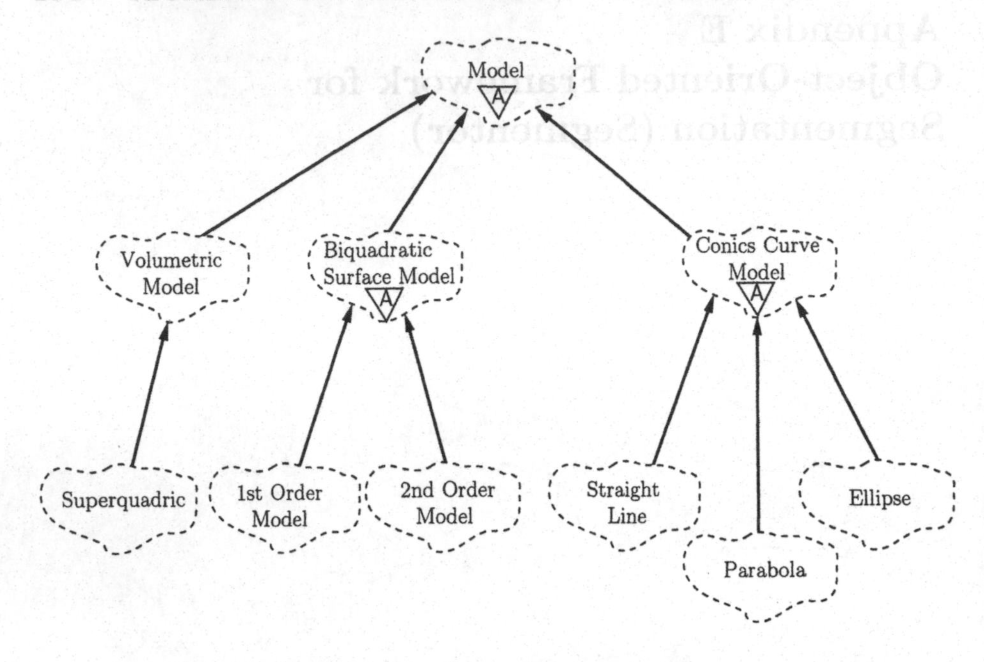

Figure E.1. Class diagram of *models*

Object-Oriented Analysis of the Recover-and-Select Paradigm

In this section we describe the main classes of the problem domain, at the highest level of abstraction, together with the examples of concrete classes for particular image domains. The examples naturally suggest the class hierarchy. Also the dynamic behavior of description objects is presented. It is expressed in terms of abstract classes, suggesting the use of polymorphism for concrete class objects. The class diagrams are presented in Booch notation (Figs. E.1 and E.2) (Booch, 1994).

World is a finite set of points from a finite dimensional vector space, with defined distance function $d(x, y)$ between the points. The distance function of a point from the subset V of the world is defined as

$$d(x, V) = \min_{y \in V} d(x, y). \tag{E.1}$$

The ε neighborhood of the subset V is defined as a set of points, such that the distance of a point from the subset is less or equal to ε, and the points are not members of the subset V

$$N_\varepsilon(V) = \{x : d(x, V) \leq \varepsilon \wedge x \notin V\}. \tag{E.2}$$

Model is an abstract entity with a defined distance $d(M, x)$ between the model and the point from the world. The distance between the model

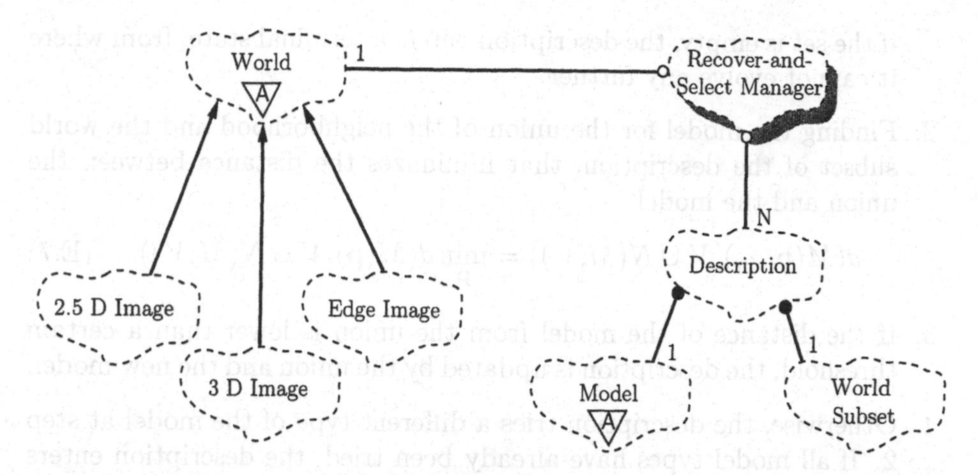

Figure E.2. Class diagram of *world, description,* and *recover-and-select class utility*

and the subset V of the world is, in turn defined as

$$d(M, V) = \sum_{x \in V} d(M, x). \tag{E.3}$$

The model is usually parameterized with a finite number of parameters. We denote the parametric model as $M(\mathbf{p})$, where $\mathbf{p}$ is a finite dimensional vector of parameters with dimension $\dim(\mathbf{p})$. To use the model in the recover-and-select paradigm one must be able to find the parameters $\mathbf{p}_{min}$ that minimize the distance between the model and the given subset V of the world

$$d(M(\mathbf{p}_{min}), V) = \min_{\mathbf{p}} d(M(\mathbf{p}), V). \tag{E.4}$$

Models do not exist in isolation from the world. As their name implies, they model the subsets of the world.

Description is a pair (V, M) of the subset of the world and the model that describes this subset. The quality of the subset representation is measured by an average description error $\bar{\xi}$, which equals the average distance between the model and the subset

$$\bar{\xi} = \frac{1}{|V|} d(M, V). \tag{E.5}$$

Descriptions are spatio-temporal dynamic objects. They evolve over time and space by

1. Forming the neighborhood of the description,

$$N(M, V) = \{x : x \in N_\varepsilon \wedge d(M, x) \leq \gamma\} \tag{E.6}$$

if the set is empty, the description enters a terminal state, from where it cannot evolve any further.

2. Finding the model for the union of the neighborhood and the world subset of the description, that minimizes the distance between the union and the model

$$d(M(\mathbf{p}_{min}), V \cup N(M,V)) = \min_{\mathbf{p}} d(M(\mathbf{p}), V \cup N(M,V)). \quad \text{(E.7)}$$

3. If the distance of the model from the union is lower than a certain threshold, the description is updated by the union and the new model.

4. Otherwise, the description tries a different type of the model at step 2. If all model types have already been tried, the description enters the terminal state and cannot evolve any further.

Recover-and-select paradigm initiates a set of descriptions in the world, and then allows them to evolve in time and space. Since the evolution of each description is computationally expensive, the description that does not model the world well enough in cooperation with other descriptions is not allowed to evolve any further, and is discarded from the overall description of the world. The selection of individual descriptions is based on the solution of the Quadratic Boolean Problem as described in Chapter 5.

User Interface

In this section we describe the Segmentor user interface (Fig. E.3) through a usage scenario used for adaptation of the paradigm to superquadric models. First, the superquadric class was derived from the abstract model class and the pure virtual functions were implemented. The growing algorithm was then tested by manually placing seeds on different parts of the image and and by observing the growth of those seeds. The maximum point distance and the maximum average description error thresholds were experimentally tested.

We observed that in some cases the models stopped growing due to our simple growing strategy, which always estimated the initial superquadric parameters for the non-linear iterative minimization from the data, not using the recovered model parameters from the previous step. This was then alleviated by recovering two models, one from the initial parameters, derived directly from the data, and the other one, recovered from the parameters of the model from the previous iteration. Once we had the basic growing algorithm working, we proceeded to automatic grid like placement of seeds. After which appropriate constants for the selection procedure were determined.

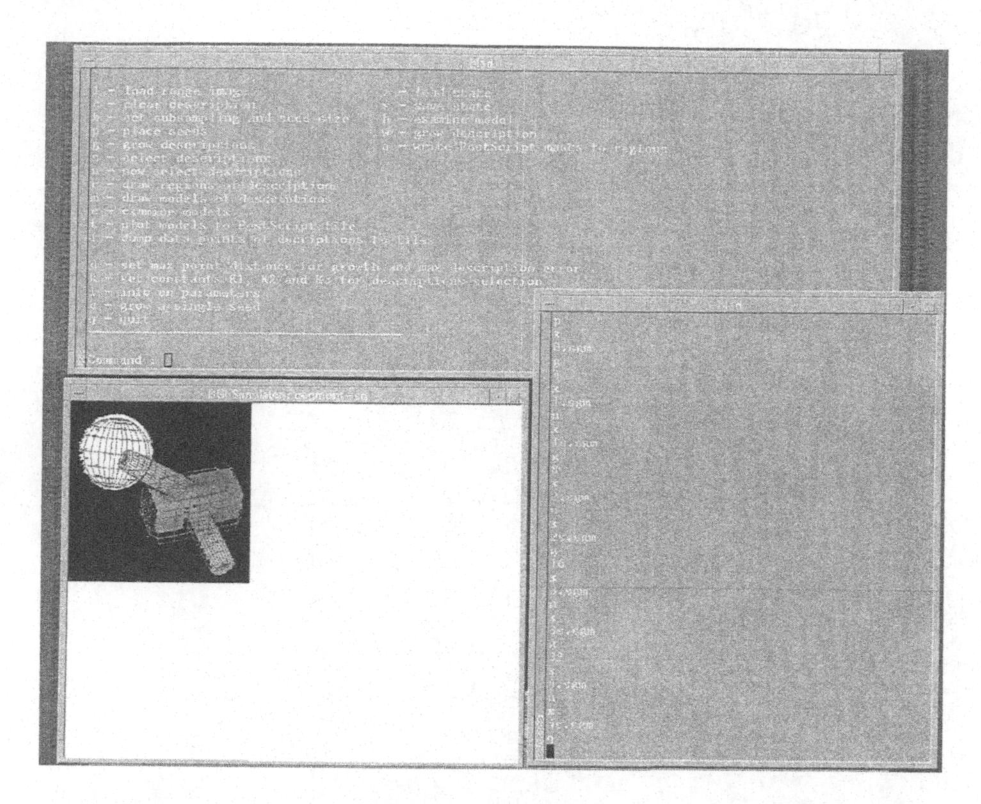

Figure E.3. A screen snapshot of a Segmentor session: the topmost window displays a list of commands and the command prompt, the left window at the bottom displays a range image and the superquadric models, the right bottom window displays a simple script file containing a sequence of commands for description growing, storage of intermediate results and selection.

An option to save and restore models from files is instrumental in this experimentation to decouple the selection phase from the growing phase. This part of the experimentation involves frequent user interaction with the models to examine their parameters.

After having a few test images successfully segmented, we can proceed to larger sets of experiments performed in a batch mode. To support such experimentation, the crucial decision is to provide a set of orthogonal commands and a simple terminal input. The segmentation is performed as a sequence of basic commands, which together with the terminal input and standard input redirection provide us with a tool for large batch experiments, where intermediate results are stored to files for later use.

Figure E.3. A screen snapshot of a Segmentor session: the topmost window displays a list of commands and the command prompt, the left window at the bottom displays a range image and the superquadric models, the right bottom window displays a simple script file containing a sequence of commands for description growing, storage of intermediate results and selection.

An option to save and restore models from files is instrumental in this experimentation to decouple the selection phase from the growing phase. This part of the experimentation involves frequent user interaction with the models to examine their parameters.

After having a few test images successfully segmented, we can proceed to larger sets of experiments performed in a batch mode. To support such experimentation, the crucial decision is to provide a set of orthogonal commands and a simple terminal input. The segmentation is performed as a sequence of basic commands, which together with the terminal input and standard input redirection provide us with a tool for large batch experiments, where intermediate results are stored to files for later use.

References

Agba, E. I., Wong, T.-L., Huang, M. Z., and Clark, A. (1993). Objects interaction using superquadrics for telemanipulation system simulation. *Journal of Robotic Systems*, 10(1):1–22.

Allen, P. K. and Michelman, P. (1990). Acquisition and interpretation of 3-D sensor data from touch. *IEEE Transactions on Robotics and Automation*, 6(4):397–404.

Arbel, T., Whaite, P., and Ferrie, F. P. (1994). Recognizing volumetric objects in the presence of uncertainty. In *Proceedings 12th IAPR International Conference on Pattern Recognition*, volume I, pages 470–476, Jerusalem, Israel. IEEE CS.

Ardizzone, E., Palazzo, M. A., and Sorbello, F. (1989). Computer reconstruction and description of 3-D objects. In Hamza, M. H., editor, *Proceedings Seventeenth IASTED International Symposium Simulation and Modelling*, pages 139–145, Lugano, Switzerland. Acta Press.

Ayoung-Chee, N., Dudek, G., and Ferrie, F. P. (1996). Enhanced 3D representation using a hybrid model. In *Proceedings 13th IAPR International Conference on Pattern Recognition*, volume I, pages 575–579, Vienna,Austria. IEEE.

Bajcsy, R. and Solina, F. (1987). Three dimensional object representation revisited. In *Proceedings First International Conference on Computer Vision*, pages 231–240, London. IEEE.

Bajcsy, R., Solina, F., and Gupta, A. (1990). Segmentation versus object representation – are they separable? In Jain, R. and Jain, A., editors, *Analysis and Interpretation of Range Images*, Perception Engineering, pages 207–223. Springer-Verlag, New York.

Bardinet, E., Cohen, L. D., and Ayache, N. (1994). Fitting 3-D data using superquadrics and free-form deformations. In *Proceedings 12th IAPR International Conference on Pattern Recognition*, volume I, pages 79–83, Jerusalem, Israel. IEEE CS.

Barr, A. H. (1981). Superquadrics and angle-preserving transformations. *IEEE Computer Graphics and Applications*, 1(1):11–23.

Barr, A. H. (1984). Global and local deformations of solid primitives. *Computer Graphics*, 18(3):21–30.

Bathe, K.-J. (1982). *Finite Element Procedures in Engineering Analysis.* Prentice-Hall.

Bergevin, R., Laurendeau, D., and Poussart, D. (1995). Registering range views of multipart objects. *Computer Vision and Image Understanding*, 61(1):1–16.

Besl, P. J. (1988). *Surfaces in Range Image Understanding.* Springer, New York.

Besl, P. J. and Jain, R. C. (1986). Invariant surface characteristics for 3D object recognition in range images. *Computer Vision, Graphics, and Image Processing*, 33(1):33–80.

Besl, P. J. and McKay, N. D. (1992). A method for registration of 3-D shapes. *IEEE Transactions on Pattern Analysis and Machine Intelligence*, 14(2):239–256.

Bhanu, B., Lee, S., and Ming, J. (1989). Adaptive image segmentation using a genetic algorithm. In *Proceedings DARPA Image Understanding Workshop*, pages 1043–1055. Morgan Kaufman.

Biederman, I. (1985). Human image understanding: Recent research and a theory. *Computer Vision, Graphics, and Image Processing*, 32:29–73.

Binford, T. O. (1971). Visual perception by a computer. In *Proceedings IEEE Conference on Systems and Controls*, Miami, FL.

Blais, G. and Levine, M. D. (1995). Registering multiview range data to create 3D computer objects. *IEEE Transactions on Pattern Analysis and Machine Intelligence*, 17(8):820–824.

Blanc, C. and Schlick, C. (1996). Ratioquadrics: An alternative model for superquadrics. *The Visual Computer*, 12(8):420–428.

Blinn, J. (1982). A generalization of algebraic surface drawing. *IEEE Transaction on Graphics*, 1(3):235–256.

Blum, H. (1973). Biological shape and visual science (part I). *Journal of Theoretical Biology*, 38:205–287.

Bobick, A. F. and Bolles, R. C. (1992). The representation space paradigm of concurrent evolving object descriptions. *IEEE Transactions on Pattern Analysis and Machine Intelligence*, 14(2):146–156.

Bolle, R. M. and Vemuri, B. C. (1991). On three-dimensional surface reconstruction methods. *IEEE Transactions on Pattern Analysis and Machine Intelligence*, 13(1):1–13.

Booch, G. (1994). *Object-Oriented Analysis and Design.* Benjamin/Cummings, Redwood City, CA, second edition.

Boult, T. E. and Gross, A. D. (1987). Recovery of superquadrics from depth information. In *Proceedings Spatial Reasoning and Multi-Sensor Fusion Workshop*, pages 128–137, St. Charles, IL. SPIE.

Boult, T. E. and Gross, A. D. (1988). On the recovery of superellipsoids. In *Proceedings DARPA Image Understanding Workshop*, pages 1052–1063.

Brady, M. (1983). Criteria for representation of shape. In Beck, J., Hope, B., and Rosenfeld, A., editors, *Human and Machine Vision*, Notes and Reports in Computer Science and Applied Mathematics, pages 39–84. Academic Press, New York.

Brady, M., Ponce, J., Yuille, A., and Asada, H. (1985). Describing surfaces. *Computer Vision, Graphics and Image Processing*, 32(1):1–28.

Brechbühler, C., Gerig, G., and Kübler, O. (1995). Parameterization of closed surfaces for 3-D shape description. *Computer Vision and Image Understanding*, 61(2):154–170.

Brooks, R. A. (1983). Model-based three-dimensional interpretation of two-dimensional images. *IEEE Transactions on Pattern Analysis and Machine Intelligence*, 5(2):140–150.

Callari, F. G. and Maniscalco, U. (1994). A new robust approach to image shading analysis and 3-D shape reconstruction. In *Proceedings 12th IAPR International Conference on Pattern Recognition*, volume I, pages 103–107, Jerusalem, Israel. IEEE CS.

Canning, J. (1991). Recognizing objects with variable appearances—the VAPOR system. Technical Report CAR-TR-563, CS-TR-2705, Computer Vision Laboratory, Center for Automation Research, University of Maryland, College Park, MD.

Chan, M. and Metaxas, D. (1994). Physics-based object pose and shape estimation from multiple views. In *Proceedings 12th IAPR International Conference on Pattern Recognition*, volume I, pages 326–330, Jerusalem, Israel. IEEE CS.

Chen, C. W., Huang, T. S., and Arrott, M. (1994a). Modeling, analysis, and visualization of left ventricle shape and motion by hierarchical decomposition. *IEEE Transaction on Pattern Analysis and Machine Intelligence*, 16(4):342–356.

Chen, D. S. (1989). A data-driven intermediate level feature extraction algorithm. *IEEE Transactions on Pattern Analysis and Machine Intelligence*, 11(7):749–758.

Chen, L.-H., Lin, W.-C., and Liao, H.-Y. M. (1994b). Recovery of superquadric primitive from stereo images. *Image and Vision Computing*, 12(5):285–296.

Chen, L.-H., Liu, Y.-T., and Liao, H.-Y. (1997). Similarity measure for superquadrics. *IEE Proceedings Vision, Image and Signal Processing*, 144(4):237–243.

Chen, Y. and Medioni, G. (1992). Object modelling by registration of multiple range images. *Image and Vision Computing*, 10(3):145–155.

Cootes, T., Taylor, C., Cooper, D., and Graham, J. (1992). Training models of shape from sets of examples. In *Proceedings of the British Machine Vision Conference*, pages 8–18.

Darrell, T., Sclaroff, S., and Pentland, A. P. (1990). Segmentation by minimal description. In *Proceedings 3rd International Conference on Computer Vision*, pages 112–116, Osaka, Japan. IEEE.

DeCarlo, D. and Metaxas, D. (1998). Shape evolution with structural and topological changes using blending. *IEEE Transactions on Pattern Recognition and Machine Intelligence*, 20(11):1186–1205.

Delingette, H., Hebert, M., and Ikeuchi, K. (1993). A spherical representation for the recognition of curved objects. In *Proceedings 4th International Conference on Computer Vision*, pages 103–112, Berlin, Germany. IEEE CS.

Dickinson, S. J., Pentland, A. P., and Rosenfeld, A. (1992a). 3-D shape recovery using distributed aspect matching. *IEEE Transaction on Pattern Analysis and Machine Intelligence*, 14(2):174–198.

Dickinson, S. J., Pentland, A. P., and Rosenfeld, A. (1992b). From volumes to views: An approach to 3-D object recognition. *CVGIP: Image Understanding*, 55(2):130–154.

Dorai, C., Wang, G., Jain, A. K., and Mercer, C. (1996). From images to models: Automatic 3D object model construction from multiple views. In *Proceedings 13th IAPR International Conference on Pattern Recognition*, volume I, pages 770–774, Vienna, Austria. IEEE.

Eggert, D. W., Fitzgibbon, A. W., and Fisher, R. B. (1996). Simultaneous registration of multiple range views for use in reverse engineering. In *Proceedings 13th IAPR International Conference on Pattern Recognition*, volume I, pages 243–247, Vienna, Austria. IEEE.

Essa, I. A., Sclaroff, S., and Pentland, A. P. (1992). A unified approach for physical and geometric modeling. *Computer Graphics Forum, The International Journal of the Eurographics Association*, 2(3):129–138.

Essa, I. A., Sclaroff, S., and Pentland, A. P. (1993). Physically-based modeling for graphics and vision. In Martin, R., editor, *Directions in Geometric Computing*. Information Geometers, U.K.

Fan, T. J. (1989). *Describing and Recognizing 3-D Objects Using Surface Properties*. Springer.

Faugeras, O. D. and Hebert, M. (1986). The representation, recognition, and locating of 3-D objects. *The International Journal of Robotics Research*, 5(3):27–52.

Faux, I. D. and Pratt, M. J. (1985). *Computational Geometry for Design and Manufacture.* Ellis Horwood, Chichester.

Feldman, J. A. (1982). Dynamic connections in neural networks. *Biological Cybernetics*, 46:27–39.

Ferrie, F. P., Lagarde, J., and Whaite, P. (1993). Darboux frames, snakes, and super-quadrics: Geometry from the bottom up. *IEEE Transactions on Pattern Analysis and Machine Intelligence*, 15(8):771–784.

Fitzgibbon, A. W., Pilu, M., and Fisher, R. B. (1996). Direct least squares fitting of ellipses. In *Proceedings 13th IAPR International Conference on Pattern Recognition*, volume I, pages 253–257, Vienna, Austria. IEEE.

Flanagan, D. L. and Hefner, O. V. (1967). Surface moulding – new tool for the engineer. *Aeronautics and Astronautics*, pages 58–62.

Franklin, W. R. and Barr, A. H. (1981). Faster calculation of superquadric shapes. *IEEE Computer Graphics and Applications*, 1(3):41–47.

Fua, P. and Hanson, A. J. (1989). Objective functions for feature discrimination. In *Proceedings of the 11th International Joint Conference on Artificial Intelligence*, pages 1596–1602, Detroit, MI. Morgan Kaufman.

Galvez, J. M. and Canton, M. (1993). Normalization and shape recognition of three-dimensional objects by 3D moments. *Pattern Recognition*, 26(5):667–681.

Gardner, M. (1965). The superellipse: A curve that lies between the ellipse and the rectangle. *Scientific American*, 213(3):222–234.

Geman, S. and Geman, D. (1984). Stochastic relaxation, Gibbs distributions, and the Bayesian restoration of images. *IEEE Transactions on Pattern Analysis and Machine Intelligence*, 6:721–741.

Goldberg, D. E. (1989). *Genetic Algorithms in Search, Optimization and Machine Learning.* Addison-Wesley, Reading, MA.

Goldstein, H. (1980). *Classical Mechanics.* Addison-Wesley, second edition.

Gross, A. D. and Boult, T. E. (1988). Error of fit measures for recovering parametric solids. In *Proceedings 2nd International Conference on Computer Vision*, pages 690–694, Tampa, FL. IEEE.

Gupta, A. and Bajcsy, R. (1990). Part description and segmentation using contour, surface and volumetric primitives. In *Proceedings Conference on Sensing and Reconstruction of 3D Objects and Scenes*, pages 203–214, Santa Clara, CA. SPIE.

Gupta, A. and Bajcsy, R. (1993). Volumetric segmentation of range images of 3D objects using superquadric models. *CVGIP: Image Understanding*, 58(3):302–326.

Gupta, A., Bogoni, L., and Bajcsy, R. (1989a). Quantitative and qualitative measures for the evaluation of the superquadric models. In *Proceedings IEEE Workshop on Interpretation of 3D Scenes*, pages 162–169, Austin, TX.

Gupta, A., Funka-Lea, G., and Wohn, K. (1989b). Segmentation, modeling and classification of the compact objects in a pile. In *Proceedings Conference on Intelligent Robots and Computer Vision VIII: Algorithms and Techniques*, volume 1192, pages 98–108, Philadelphia, PA. SPIE.

Gupta, A. and Liang, C.-C. (1993). 3-D model-data correspondence and nonrigid deformation. In *Proceedings Computer Vision and Pattern Recognition Conference*, pages 756–757, New York, NY. IEEE.

Gupta, A., O'Donnell, T., and Singh, A. (1994). Segmentation and tracking of cine cardiac MR and CT images using a 3-D deformable model. In *Proceedings of Conference on Computers in Cardiology*, pages 661–664. IEEE.

Hager, G. D. (1994). Task-directed computation of qualitative decisions from sensor data. *IEEE Transactions on Robotics and Automation*, 10(4):415–429.

Han, S., Goldgof, D. B., and Bowyer, K. W. (1993). Using hyperquadrics for shape recovery from range data. In *Proceedings 4th International Conference on Computer Vision*, pages 492–496, Berlin, Germany. IEEE CS.

Hanson, A. J. (1988). Hyperquadrics: Smoothly deformable shapes with convex polyhedral bounds. *Computer Vision, Graphics, and Image Processing*, 44:191–210.

Hildebrand, F. B. (1987). *Introduction to Numerical Analysis.* Dover, New York, NY.

Hochberg, J. (1981). Levels of perceptual organization. In Kubovy, M. and Pomerantz, J. R., editors, *Perceptual Organization*, pages 255–276. Lawrence Erlbaum Associates, New Jersey.

Hoffman, D. D. and Richards, W. A. (1985). Parts of recognition. *Cognition*, 18:65–96.

Hoover, A., Jean-Baptiste, G., Jiang, X., Flynn, P. J., Bunke, H., Goldgof, D. B., Bowyer, K., Eggert, D. W., Fitzgibbon, A., and Fisher, R. B. (1996). An experimental comparison of range image segmentation algorithms. *IEEE Transactions on Pattern Analysis and Machine Intelligence*, 18(7):673–689.

Hopfield, J. J. and Tank, D. W. (1985). "Neural" computation of decisions in optimization problems. *Biological Cybernetics*, 52:141–152.

Hopfield, J. J. and Tank, D. W. (1986). Computing with neural circuits: A model. *Science*, 233:625–633.

Horikoshi, T. and Kasahara, H. (1990). 3-D shape indexing language. In *Proceedings International Conference on Computers and Communications*, pages 493–499, Scottsdale, AZ. IEEE.

Horikoshi, T. and Suzuki, S. (1993). 3D parts decomposition from sparse range data using information criterion. In *Proceedings Computer Vision and Pattern Recognition Conference*, pages 168–173, New York, NY. IEEE.

Horn, B. K. P. (1986). *Robot Vision.* MIT Press, Cambridge, MA.

Ikeuchi, K. and Hebert, M. (1996). Task-oriented vision. In Landy, M. S., Maloney, L. T., and Pavel, M., editors, *Exploratory Vision, The Active Eye*, Perception Engineering, pages 257–277. Springer, New York.

Jaklič, A. (1996). Segmentor: An object-oriented framework for image segmentation. Technical Report LRV-96-2, Computer Vision Laboratory, Faculty of Computer and Information Science, University of Ljubljana.

Jaklič, A. (1997). *Construction of CAD models from range images.* PhD thesis, University of Ljubljana, Faculty of Computer and Information Science.

Jaklič, A. and Solina, F. (1994). Construction of CAD models for reverse engineering. In Zajc, B. and Solina, F., editors, *Proceedings Third Electrotechnical and Computer Science Conference ERK'94*, volume B, pages 251–254, Portorož, Slovenia. Slovenia Section IEEE.

Jojić, N. and Huang, T. S. (2000). Computer vision and graphics techniques for modeling dressed humans. In Leonardis, A., Solina, F., and Bajcsy, R., editors, *The confluence of computer vision and computer graphics*, pages 179–200. Kluwer, Dordrecht.

Jouvencel, B. and Simphor, J. E. (1991). The variable modelling of mobile robot environments. In *Proceedings International Workshop on Intelligent Robots and Systems*, pages 1044–1051, Osaka, Japan. IEEE/RSJ.

Kang, S. B. and Ikeuchi, K. (1993). The complex EGI: A new representation for 3-D pose determination. *IEEE Transactions on Pattern Analysis and Machine Intelligence*, 15(7):707–721.

Keeler, K. C. (1991). Map representations and optimal encoding for image segmentation. Technical Report CICS-TH-292, Center for Intelligent Control Systems.

Kender, J. R. and Freudenstein, D. (1987). What is a "degenerate view"? In *Proceedings DARPA Image Understanding Workshop*, pages 589–598, Los Angeles.

Keren, D., Cooper, D., and Subrahmonia, J. (1994). Describing complicated objects by implicit polynomials. *IEEE Transactions on Pattern Analysis and Machine Intelligence*, 16(1):38–53.

Khatib, O. (1986). Real-time obstacle avoidance for manipulators and mobile robots. *International Journal of Robotic Research*, 5(1):90–98.

Kirkpatrick, S., Gelatt, C. D., and Vecchi, M. P. (1983). Optimization by simulated annealing. *Science*, 220:671–680.

Koch, C. and Ullman, S. (1984). Selecting one among many: A simple network implementing shifts in selective visual attention. Technical Report A.I. Memo 770, C.B.I.P. Paper 3, Artificial Intelligence Laboratory, Massachusetts Institute of Technology.

Koenderink, J. J. (1990). *Solid Shape.* MIT Press, Cambridge, MA.

Koenderink, J. J. and van Doorn, A. (1979). The internal representation of solid shape with respect to vision. *Biological Cybernetics*, 32:211–216.

Koenderink, J. J. and van Doorn, A. (1982). The shape of smooth objects and the way contours end. *Perception*, 11:129–137.

Krotkov, E. P. (1989). *Active Computer Vision by Cooperative Focus and Stereo.* Springer, New York.

Kumar, S. and Goldgof, D. (1994). A robust technique for the estimation of the deformable hyperquadrics from images. In *Proceedings 12th IAPR International Conference on Pattern Recognition*, pages 74–78, Jerusalem, Israel. IEEE CS.

Leclerc, Y. G. (1989a). Constructing simple stable descriptions for image partitioning. *International Journal of Computer Vision*, 3:73–102.

Leclerc, Y. G. (1989b). *The Local Structure of Image Intensity Discontinuities.* PhD thesis, Computer Vision and Robotics Laboratory, Department of Electrical Engineering, McGill University, Montreal, Canada.

Leonardis, A. (1993). *Image analysis using parametric models: Model-recovery and model-selection paradigm.* PhD thesis, University of Ljubljana, Faculty of Electrical Engineering and Computer Science.

Leonardis, A., Gupta, A., and Bajcsy, R. (1990). Segmentation as the search for the best description of the image in terms of primitives. In *Proceedings 3rd International Conference on Computer Vision*, pages 121–125, Osaka, Japan. IEEE.

Leonardis, A., Gupta, A., and Bajcsy, R. (1995). Segmentation of range images as the search for geometric parametric models. *International Journal of Computer Vision*, 14:253–277.

Leonardis, A., Jaklič, A., Kverh, B., and Solina, F. (1996). Simultaneous recovery of surface and superquadric models. In Pinz, A., editor, *Pattern Recognition 1996, Proceedings of 20. Workshop of the Austrian Pattern Recognition Group (ÖAGM/AAPR), SchloßSeggau, Leibnitz, May, 1996*, volume 90 of *Shriftenreihe der ÖCG*, pages 27–36. R. Oldenburg, Wien.

Leonardis, A., Jaklič, A., and Solina, F. (1997). Superquadrics for segmentation and modeling range data. *IEEE Transactions on Pattern Recognition and Machine Intelligence*, 19(11):1289–1295.

Leyton, M. (1988). A process-grammar for shape. *Artificial Intelligence*, 34:213–247.

Liang, C.-C., Lin, W.-C., and Chen, C.-T. (1992). Deformation process modeling in medical imaging. In *Proceedings International Conference on Systems, Man, and Cybernetics*, pages 1358–1363. IEEE.

Löffelmann, H. and Gröller, E. (1994). Parameterizing superquadrics. Technical Report TR-186-2-94-05, Institute of Computer Graphics, Vienna University of Technology, Karlsplatz 13/186/2, A-1040 Vienna, Austria.

Loria, G. (1910). *Spezielle algebraische und transzendente ebene Kurven.* B. G. Teubner, Leipzig, Berlin.

Lowe, D. (1985). *Perceptual Organization and Visual Recognition.* Kluwer, Boston, MA.

Mallat, S. G. (1989). A theory of multi-resolution signal decomposition: The wavelet representation. *IEEE Transactions on Pattern Analysis and Machine Intelligence*, 11(7):674–693.

Marr, D. (1982). *Vision, a Computational Investigation into the Human Representation and Processing of Visual Information.* Freeman, San Francisco, CA.

Marr, D. and Nishihara, K. (1978). Representation and recognition of the spatial organization of three-dimensional shapes. *Proceedings of the Royal Society, London*, B 200:269–294.

Martin, R. R. and Várady, T., editors (1998). *Report on merging and applications, with further contributions on basic geometry and geometric model creation*, number 1998/4 in GML, Budapest, Hungary. Computer and Automation Institute, Hungarian Academy of Sciences.

Marzani, F., Maliet, Y., Legrand, L., and Dusserre, L. (1997). A computer model based on superquadrics for the analysis of movement disabilities. In *Proceedings 19th International Conference of the IEEE Engineering in Medicine and Biology Society*, pages 1817–1820, Chicago, IL.

Maver, J. and Bajcsy, R. (1993). Occlusions as a guide for planning the next view. *IEEE Transaction on Pattern Analysis and Machine Intelligence*, 15(5):417–433.

McCarthy, L., Stiles, R., Pontecorvo, M., and Grant, F. (1993). Spatial considerations for instructional development in a virtual environment. In *Conference on Intelligent Computer-Aided Training and Virtual Environment Technology*, NASA Johnson Space Center, Houston, TX.

McInerney, T. and Terzopoulos, D. (1993). A finite element model for 3D shape reconstruction and nonrigid motion tracking. In *Proceedings 4th International Conference on Computer Vision*, pages 518–523, Berlin, Germany. IEEE.

Metaxas, D. and Dickinson, S. J. (1993). Integration of quantitative and qualitative techniques for deformable model fitting from orthographic, perspective, and stereo projections. In *Proceedings 4th International Conference on Computer Vision*, pages 641–649, Berlin, Germany. IEEE.

Metaxas, D., Koh, E., and Badler, N. I. (1997). Multi-level shape representation using global deformations and locally adaptive finite elements. *International Journal of Computer Vision*, 25(1):49–61.

Metaxas, D. and Terzopoulos, D. (1991). Constrained deformable superquadrics and nonrigid motion tracking. In *Proceedings Computer Vision and Pattern Recognition Conference*, pages 337–343. IEEE.

Metaxas, D. and Terzopoulos, D. (1992). Dynamic deformation of solid primitives with constraints. *Computer Graphics*, 26(2):309–311.

Metaxas, D. and Terzopoulos, D. (1993). Shape and nonrigid motion estimation through physics-based synthesis. *IEEE Transactions on Pattern Analysis and machine Intelligence*, 15(6):580–591.

Mohan, R. and Nevatia, R. (1989). Using perceptual organization to extract 3-D structures. *IEEE Transactions and Pattern Analysis and Machine Intelligence*, 11(11):1121–1139.

Montiel, M. E., Aguado, A. S., and Zaluska, E. J. (1998). Surface subdivision for generating superquadrics. *The Visual Computer*, 14(1):1–17.

Muraki, S. (1991). Volumetric shape description of range data using "blobby model". *Computer graphics*, 25(4):227–235.

Nackman, L. R. and Pizer, S. M. (1985). Three dimensional shape description using the symmetric axis transform I: Theory. *IEEE Transactions and Pattern Analysis and Machine Intelligence*, 7(2):187–202.

Nevatia, R. and Binford, T. (1977). Description and recognition of curved objects. *Artificial Intelligence*, 8:77–98.

O'Donnell, T., Fang, X.-S., Boult, T. E., and Gupta, A. (1994). The extruded generalized cylinder: A deformable model for object recovery.

In *Proceedings Computer Vision and Pattern Recognition Conference*, pages 174–181, Seattle, WA. IEEE.

O'Donnell, T., Gupta, A., and Boult, T. E. (1995). The hybrid volumetric ventriculoid: A model for MR-SPAMM 3-D analysis. In *Proceedings Computers in Cardiology*, pages 5–8, Vienna, Austria. IEEE.

Park, J., Metaxas, D., and Axel, L. (1995). Volumetric deformable models with parameter functions: A new approach to the 3D motion analysis of the LV from MRI-SPAMM. In *Proceedings of International Conference on Computer Vision*, pages 700–705, Boston, MA. IEEE.

Park, J., Metaxas, D., and Young, A. A. (1994). Deformable models with parameter functions: Application to heart-wall modeling. In *Proceedings Computer Vision and Pattern Recognition Conference*, pages 437–442, Seattle, WA. IEEE.

Paul, R. P. (1981). *Robor Manipulators: Mathematics, Programming, and Control.* MIT Press, Cambridge, MA.

Pentland, A. P. (1986). Perceptual organization and the representation of natural form. *Artificial Intelligence*, 28(2):293–331.

Pentland, A. P. (1987). Recognition by parts. In *Proceedings First International Conference on Computer Vision*, pages 612–620, London, England. IEEE.

Pentland, A. P. (1989a). Part segmentation for object recognition. *Neural Computation*, 1(1):82–91.

Pentland, A. P. (1989b). Thingworld: A multibody simulation system with low computational complexity. In Sriram, D. and Logcher, R., editors, *Cooperative Computer Aided Design*, volume 492 of *Lecture Notes in Computer Science*, pages 560–583. Springer, Berlin.

Pentland, A. P. (1990). Automatic extraction of deformable part models. *International Journal of Computer Vision*, 4:107–126.

Pentland, A. P., Horowitz, B., and Sclaroff, S. (1991). Non-rigid motion and structure from contour. In *IEEE Workshop on Visual Motion*, pages 288–293, Princeton, NJ.

Pentland, A. P. and Sclaroff, S. (1991). Closed-form solutions for physically based shape modeling and recognition. *IEEE Transactions on Pattern Analysis and Machine Intelligence*, 13(7):715–729.

Pentland, A. P. and Williams, J. (1989a). Good vibrations: Modal dynamics for graphics and animation. *Computer Graphics*, 23(3):215–222.

Pentland, A. P. and Williams, J. (1989b). Perception of non-rigid motion: Inference of shape, material, and force. In *Proceedings International Joint Conference on Artificial Intelligence*, volume 2, pages 1565–1570, Detroit, MI.

Pentland, A. P. and Williams, J. (1989c). Virtual manufacturing. In *Proceedings NSF Conference on Manufacturing Design*, pages 301–316, Amherst, MA.

Pentland, A. P., Williams, J., and Connors, J. (1989). Interactive integrated design — visualization of form and process. In *Proceedings 3rd International Conference on Human-Computer Interactions*, pages 18–22, Boston, MA.

Pilu, M. and Fisher, R. B. (1995). Equal-distance sampling of superellipse models. In *Proceedings of the British Machine Vision Conference*, volume I, pages 257–266, Birmingham.

Pilu, M. and Fisher, R. B. (1996a). Part segmentation from 2D edge images by the MDL criterion. In *Proceedings of the British Machine Vision Conference*, pages 83–92, Edinburgh.

Pilu, M. and Fisher, R. B. (1996b). Recognition of geons by parametric deformable contour models. In Buxton, B. and Cipolla, R., editors, *Computer Vision – ECCV'96, Cambridge, UK, Vol. I*, volume 1064 of *Lecture Notes in Computer Science*, pages 71–82, Berlin. Springer.

Pilu, M., Fitzgibbon, A. W., and Fisher, R. B. (1996). Training PDMs on models: The case of deformable superellipses. In *Proceedings of the British Machine Vision Conference*, pages 373–382, Edinburgh.

Pomerantz, J. R. (1981). Perceptual organization in information processing. In Kubovy, M. and Pomerantz, J. R., editors, *Perceptual organization*, pages 141–179. Lawrence Erlbaum Associates, New Jersey.

Pomerantz, J. R. and Kubovy, M. (1981). Perceptual organization: An overview. In Kubovy, M. and Pomerantz, J. R., editors, *Perceptual Organization*, pages 423–456. Lawrence Erlbaum Associates, New Jersey.

Ponce, J., Chelberg, D., and Mann, W. B. (1989). Invariant properties of straight homogeneous generalized cylinders and their contours. *IEEE Transactions and Pattern Analysis and Machine Intelligence*, 11(9):951–966.

Press, W. H., Flannery, B. P., Teukolsky, S. A., , and Vettering, W. T. (1986). *Numerical Recipes*. Cambridge University Press, Cambridge.

Raja, N. S. and Jain, A. K. (1992). Recognizing geons from superquadrics fitted to range data. *Image and Vision Computing*, 10(3):179–190.

Raja, N. S. and Jain, A. K. (1994). Obtaining generic parts from range images using a multi-view representation. *CVGIP: Image Understanding*, 60(1):44–64.

Rao, K. and Nevatia, R. (1988). Computing volume descriptions from sparse 3-D data. *International Journal of Computer Vision*, 2(1):33–50.

Ripley, C. S., Beals, S. P., Joganic, E. F., Pomatto, J., Manwaring, K., Moss, S. D., Sommer, H. J., Eckhardt, R. B., and Eckhardt, J. T. (1995). Mathematical modeling of cranial molding: Superquadric images document rapid change of cranial form in response to environmental influences. *American Journal of Physiological Anthropology*, Supplement 20:89.

Rissanen, J. (1983). A universal prior for the integers and estimation by minimum description length. *Annals of Statistics*, 11(2):416–431.

Rissanen, J. (1984). Universal coding, information, prediction, and estimation. *IEEE Transactions on Information Theory*, 30:629–636.

Rivlin, E., Dickinson, S. J., and Rosenfeld, A. (1994). Recognition by functional parts. In *Proceedings Computer Vision and Pattern Recognition Conference*, pages 267–274, Seattle, WA. IEEE.

Rock, I. (1983). *The Logic of Perception*. MIT Press, Cambridge, MA.

Rosenfeld, A. and Kak, A. (1982). *Digital Picture Processing*. Academic Press, Orlando, FL.

Saito, H. and Kimura, M. (1996). Superquadrics modeling of multiple objects from shading images using genetic algorithms. In *Proceedings 22th International Conference on Industrial Electronics, Control, and Instrumentation*, pages 1589–1593, Taipei.

Saito, H. and Tsunashima, N. (1994). Estimation of 3-D parametric models from shading image using genetic algorithms. In *Proceedings 12th IAPR International Conference on Pattern Recognition*, pages 668–670, Jerusalem, Israel. IEEE CS.

Scales, L. E. (1985). *Introduction to Non-Linear Optimization*. Springer, New York.

Schudy, R. B. and Ballard, D. H. (1979). Towards an anatomical model of hearth motion as seen in 4-D cardiac ultrasound data. In *Proceedings 6th Conference on Computer Applications in Radiology and Computer-Aided Analysis of Radiological Images*.

Sclaroff, S. and Pentland, A. P. (1991). Generalized implicit functions for computer graphics. *Computer Graphics*, 25(4):247–250.

Sclaroff, S. and Pentland, A. P. (1995). Modal matching for correspondence and recognition. *IEEE Transactions on Pattern Analysis and Machine Intelligence*, 17(6):545–561.

Sederberg, T. W. and Parry, S. R. (1986). Free-form deformation of solid geometric models. *Computer Graphics*, 20(4):151–160.

Shannon, C. (1948). A mathematical theory of communication. *Bell Systems Technical Journal*, 27:379–423.

Shiang, T. Y., Sommer, H. J., and Eckhardt, R. B. (1993). Quantification of cranial surface asymmetry using superquadrics. *Biomedical Engineering — Applications, Basis & Communications*, 5(1):20–37.

Skočaj, D. (1999). Automatic modeling of 3-dimensional colored objects using range sensor. Master's thesis, University of Ljubljana, Faculty of Computer and Information Science, Slovenia.

Skočaj, D. and Leonardis, A. (2000). Acquiring range images of objects with non-uniform reflectance using high dynamic scale radiance maps. In Leonardis, A., Solina, F., and Bajcsy, R., editors, *The Confluence of Computer Vision and Computer Graphics*, pages 105–122. Kluwer, Dordrecht.

Solina, F. (1987). *Shape recovery and segmentation with deformable part models*. PhD thesis, University of Pennsylvania, Department of Computer and Information Science.

Solina, F. and Bajcsy, R. (1986). Shape and function. In *Proceedings Intelligent Robots and Computer Vision*, Vol. 726, pages 284–291, Cambridge, MA. SPIE.

Solina, F. and Bajcsy, R. (1987). Range image interpretation of mail pieces with superquadrics. In *AAAI-87, Proceedings Sixth National Conference on Artificial Intelligence*, volume 2, pages 733–737, Seattle, Washington. Morgan Kaufmann.

Solina, F. and Bajcsy, R. (1989). Recovery of mail piece shape from range images using 3-D deformable models. *International Journal of Research & Engineering, Postal Applications*, Inaugural Issue:125–131.

Solina, F. and Bajcsy, R. (1990). Recovery of parametric models from range images: The case for superquadrics with global deformations. *IEEE Transactions on Pattern Analysis and Machine Intelligence*, 12(2):131–147.

Solina, F. and Leonardis, A. (1995). Shape decomposition using part-models of different granularity. In Arcelli, C., Cordella, L. P., and di Baja, G. S., editors, *Aspects of Visual Form Processing*, pages 522–531. World Scientific, Singapore.

Solina, F. and Leonardis, A. (1998). Proper scale for modeling visual data. *Image and Vision Computing*, 16(2):89–98.

Solina, F., Leonardis, A., Jaklič, A., and Kverh, B. (1998). Reverse engineering by means of range image interpretation. In Kopacek, P. and Noe, D., editors, *Intelligent Assembly and Disassembly, A Proceedings volume from the IFAC workshop, Bled, Slovenia, 21–23 May 1998*, pages 153–158, Oxford, UK. Pergamon.

Soucy, M. and Laurendeau, D. (1995). A general surface approach to the integration of a set of range views. *IEEE Transactions on Pattern Analysis and Machine Intelligence*, 17(4):344–358.

Stark, L. (1994). Recognizing object function through reasoning about 3-D shape and dynamic physical properties. In *Proceedings Computer*

Vision and Pattern Recognition Conference, pages 546–553, Seattle, WA. IEEE.

Stark, L. and Bowyer, K. (1991). Achieving generalized object recognition through reasoning about association of function to structure. *IEEE Transactions on Pattern Analysis and Machine Intelligence*, 13(10):1097–1104.

Strang, G. (1988). *Linear Algebra and its Applications*. Harcourt Brace Jovanovich, third edition.

Sullivan, S., Sandford, L., and Ponce, J. (1994). Using geometric distance fits for 3-D object modeling and recognition. *IEEE Transactions on Pattern Analysis and Machine Intelligence*, 16(12):1183–1196.

Tasdizen, T., Tarel, J.-P., and Cooper, D. B. (1999). Algebraic curves that work better. In *Proceedings 1999 IEEE Computer Society Conference on Computer Vision and Pattern Recognition*, volume II, pages 35–41, Santa Barbara, CA.

Taubin, G. (1991). Estimation of planar curves, surfaces, and nonplanar space curves defined by implicit equations with applications to edge and range image segmentation. *IEEE Transactions on Pattern Analysis and Machine Intelligence*, 13(11):1115–1138.

Taubin, G., Cukierman, F., Sullivan, S., Ponce, J., and Kriegman, D. J. (1994). Parameterized families of polynomials for bounded algebraic curve and surface fitting. *IEEE Transactions on Pattern Analysis and Machine Intelligence*, 16(3):287–303.

Terzopoulos, D. and Metaxas, D. (1991). Dynamic 3D models with local and global deformations: Deformable superquadrics. *IEEE Transactions on Pattern Analysis and Machine Intelligence*, 13(7):703–714.

Terzopoulos, D., Witkin, A., and Kass, M. (1988a). Constraints on deformable models: Recovering 3D shape and nonrigid motion. *Artificial Intelligence*, pages 91–123.

Terzopoulos, D., Witkin, A., and Kass, M. (1988b). Symmetry-seeking models for 3D object reconstruction. *International Journal of Computer Vision*, 1(3):211–221.

Thompson, D. (1961 (First publ. 1917)). *On Growth and Form*. Cambridge University Press, Cambridge.

Tijerino, Y. A., Abe, S., Miyasato, T., and Kishino, F. (1994). What you say is what you see — interactive generation, manipulation and modification of 3-D shapes based on verbal descriptions. *Artificial Intelligence Review*, 8:215–234.

Turk, G. and Levoy, M. (1994). Zippered polygon meshes from range images. In *SIGGRAPH'94 Computer Graphics Proceedings, Annual Conference Series*, pages 311–247.

Tversky, B. and Hemenway, K. (1984). Objects, parts, and categories. *Journal of Experimental Psychology: General*, 113(2):169–193.

Ulupinar, F. and Nevatia, R. (1993). Perception of 3-D surfaces from 2-D contours. *IEEE Transactions on Pattern Analysis and Machine Intelligence*, 15(1):3–18.

van Dop, E. R. and Regtien, P. P. L. (1998). Fitting undeformed superquadrics to range data: Improving model recovery and classification. In *Proceedings 1998 IEEE Computer Society Conference on Computer Vision and Pattern Recognition*, pages 396–401, Santa Barbara, CA.

Vemuri, B. C. and Radisavljevic, A. (1993). From global to local, a continuum of shape models with fractal priors. In *Proceedings Computer Vision and Pattern Recognition Conference*, pages 307–313, New York, NY. IEEE.

Vidmar, A. and Solina, F. (1992). Reconstruction of superquadrics from 2d contours. In Klette, R. and Kropatsch, W., editors, *Theoretical Foundations of Computer Vision*, Mathematical Research, Vol. 69, pages 227–240. Akademie Verlag, Berlin.

Volpe, R. and Khosla, P. (1990). Manipulator control with superquadric artificial potential functions: Theory and experiments. *IEEE Transactions on Systems, Man, and Cybernetics*, 20(6):1423–1436.

Wertheimer, M. (1923). Principles of perceptual organisation. In Ellis, W. H., editor, *Source Book of Gestalt Psychology*. London, New York.

Whaite, P. and Ferrie, F. P. (1991). From uncertainty to visual exploration. *IEEE Transactions on Pattern Analysis and Machine Intelligence*, 13(10):1038–1049.

Whaite, P. and Ferrie, F. P. (1993). Active exploration: Knowing when we're wrong. In *Proceedings 4th International Conference on Computer Vision*, pages 41–48, Berlin, Germany. IEEE CS.

Whaite, P. and Ferrie, F. P. (1997). Autonomous exploration: Driven by uncertainty. *IEEE Transactions on Pattern Analysis and Machine Intelligence*, 19(3):193–205.

Winston, P., Binford, T., Katz, B., and Lowry, M. (1983). Learning physical descriptions from functional definitions, examples, and precedents. In *AAAI'83, Proceedings National Conference on Artificial Intelligence*, pages 433–439.

Witkin, A. P., Fleischer, K., and Barr, A. (1987). Energy constraints on parameterized models. *Computer Graphics*, 21(4):225–232.

Wu, K. and Levine, M. D. (1994a). Recovering parametric geons from multiview range data. In *Proceedings Computer Vision and Pattern Recognition Conference*, pages 159–166, Seattle, WA. IEEE.

Wu, K. and Levine, M. D. (1994b). Shape approximation: From multiview range images to parametric geons. In *Proceedings 12th IAPR In-*

ternational Conference on Pattern Recognition, volume I, pages 622–625, Jerusalem, Israel. IEEE CS.

Wuyts, T. and Eycken, L. V. (1997). Segmentation based coding for very low bitrates. In *Proceedings Conference on visual communications and image processing*, volume 3024, part II, pages 690–698, San Jose, CA.

Wyvill, B. and Wyvill, G. (1989). Field functions for implicit surfaces. *The Visual Computer*, 5:75–82.

Yokoya, N., Kaneta, M., and Yamamoto, K. (1992). Recovery of superquadric primitives from a range image using simulated annealing. In *Proceedings 11th IAPR International Conference on Pattern Recognition*, volume A, pages 168–172. IEEE Computer Society Press.

Young, A. A. and Axel, L. (1992). Three-dimensional motion and deformation of the heart wall: Estimation with spatial modulation of magnetization—a model-based approach. *Radiology*, 185(1):241–247.

Zarrugh, M. Y. (1985). Display and inertia parameters of superellipsoids as generalized constructive solid geometry primitives. In *Computers in Engineering*, pages 317–328. ASME.

Zerroug, M. and Nevatia, R. (1994). Segmentation and recovery of SHGCs from a real intensity image. In Eklundh, J.-O., editor, *Proceedings Third European Conference on Computer Vision, Volume I*, Lecture Notes in Computer Science 800, pages 319–330, Stockholm, Sweden. Springer.

Zerroug, M. and Nevatia, R. (1999). Part-based 3D descriptions of complex objects from a single image. *IEEE Transactions on Pattern Analysis and Machine Intelligence*, 21(9):835–848.

Zhou, L. and Kambhamettu, C. (1999). Extending superquadrics with exponent functions: Modeling and reconstruction. In *Proceedings 1999 IEEE Computer Society Conference on Computer Vision and Pattern Recognition*, volume II, pages 73–78.

ternational Conference on Pattern Recognition*, volume I, pages 622–625, Jerusalem, Israel. IEEE CS.

Wuyts, T. and Eycken, L. V. (1997). Segmentation based coding for very low bitrates. In *Proceedings Conference on visual communications and image processing*, volume 3024, part II, pages 690–698, San Jose, CA.

Wyvill, B. and Wyvill, G. (1989). Field functions for implicit surfaces. *The Visual Computer*, 5:75–82.

Yokoya, N., Kaneta, M., and Yamamoto, K. (1992). Recovery of superquadric primitives from a range image using simulated annealing. In *Proceedings 11th IAPR International Conference on Pattern Recognition*, volume A, pages 168–172. IEEE Computer Society Press.

Young, A. A. and Axel, L. (1992). Three-dimensional motion and deformation of the heart wall: Estimation with spatial modulation of magnetization — a model-based approach. *Radiology*, 185(1):241–247.

Zarrugh, M. Y. (1985). Display and inertia parameters of superellipsoids as generalized constructive solid geometry primitives. In *Computers in Engineering*, pages 317–328. ASME.

Zerroug, M. and Nevatia, R. (1994). Segmentation and recovery of SHGCs from a real intensity image. In Eklundh, J.-O., editor, *Proceedings Third European Conference on Computer Vision*, Volume 1, Lecture Notes in Computer Science 800, pages 319–330, Stockholm, Sweden. Springer.

Zerroug, M. and Nevatia, R. (1999). Part-based 3D descriptions of complex objects from a single image. *IEEE Transactions on Pattern Analysis and Machine Intelligence*, 21(9):835–848.

Zhou, L. and Kambhamettu, C. (1999). Extending superquadrics with exponent functions: Modeling and reconstruction. In *Proceedings 1999 IEEE Computer Society Conference on Computer Vision and Pattern Recognition*, volume II, pages 73–78.

About the Authors

Dr. Aleš Jaklič is a teaching assistant at the Faculty of Computer and Information Science of the University of Ljubljana in Ljubljana, Slovenia. He received a Dipl. Ing. (1989) in electrical engineering and M. Sc. (1992) and Ph. D. degrees (1997) in computer science, all from the University of Ljubljana.

Dr. Aleš Leonardis is an associate professor of computer and information science at the Faculty of Computer and Information Science of the University of Ljubljana in Ljubljana, Slovenia. He received Dipl. Ing. (1985) and M. Sc. (1988) degrees in electrical engineering and a Ph. D. degree (1993) in computer science, all from the University of Ljubljana. His main research interests in computer vision are robust segmentation, object recognition, and object/scene modeling.

Dr. Franc Solina is a professor of computer and information science at the Faculty of Computer and Information Science of the University of Ljubljana in Ljubljana, Slovenia where he is also the head of the Computer Vision Laboratory. He received Dipl. Ing. (1979) and M. Sc. degrees (1982) in electrical engineering from the University of Ljubljana and a Ph. D. degree (1987) in computer science from the University of Pennsylvania. His main research interests in computer vision are segmentation and part-level object representation.

Author Index

Topic Index